传统之茧

欧亚大陆民族文化中的丝绸

Cocoon of Traditions

Silk in the National Cultures of Eurasia

深圳市南山博物馆　编

文物出版社

图书在版编目（CIP）数据

传统之茧：欧亚大陆民族文化中的丝绸：汉英对照 /
深圳市南山博物馆编 . —— 北京：文物出版社，2024.
11. —— ISBN 978-7-5010-8567-5

Ⅰ . K866.92；TS145.3-64

中国国家版本馆 CIP 数据核字第 2024EL0021 号

传统之茧
欧亚大陆民族文化中的丝绸

Cocoon of Traditions
Silk in the National Cultures of Eurasia

编　　者：深圳市南山博物馆
责任编辑：智　朴
责任印制：张道奇
出版发行：文物出版社
社　　址：北京市东城区东直门内北小街 2 号楼
邮　　编：100007
网　　址：http://www.wenwu.com
邮　　箱：wenwu1957@126.com
经　　销：新华书店
装帧设计：雅昌（深圳）设计中心　冼玉梅
印　　制：雅昌文化（集团）有限公司
开　　本：889mm×1194mm　1/16
印　　张：17
版　　次：2024 年 11 月第 1 版
印　　次：2024 年 11 月第 1 次印刷
书　　号：ISBN 978-7-5010-8567-5
定　　价：390.00 元

深圳市南山博物馆馆长 致辞

　　2024 年，正值中俄建交 75 周年，在俄罗斯联邦文化部的支持下，深圳市南山博物馆与俄罗斯民族博物馆合作举办展览——"传统之茧：欧亚大陆民族文化中的丝绸"。在许多民族中，虽然存在各自的丝绸传统偏好，但它们并不排斥引进适应其社会文化需求的名贵丝绸用品和材料，并通过融合和创新最终形成了别具一格的"传统之茧"。

　　俄罗斯民族博物馆是世界上最大的民族博物馆之一，馆藏有 50 余万件展品，从 19 世纪 90 年代中期开始收集第一批藏品。俄罗斯民族博物馆通过展出工具、家庭用品、民间服饰、器皿和礼仪用品等类型的馆藏，让人们了解当地居民的职业、住所、节日、仪式和宗教信仰。

　　本次展览展出的 283 件 / 套展品主要集中在 19 到 20 世纪之间，包括中国丝绸制作的长袍、中亚手工织物样本、伊朗丝毯、东南高加索地区的"克拉加依"头巾、伏尔加河地区芬兰乌戈尔民族丰富的丝绸刺绣服装等，观众可以通过这些精美绝伦的丝织品，感受到欧亚大陆各民族文化中机织丝制品和手工丝制品的多样性。

　　"文明因交流而多彩，文明因互鉴而丰富。"本次展览是南山博物馆"一带一路"沿线地区文明主题系列展览之一，希望展览能给观众带来对于欧亚大陆这片土地无尽的遐想，充分展现出丝绸之路文明交流互通的现象。

戚鑫

深圳市南山博物馆馆长

Preface by the Director of Nanshan Museum

In 2024, the 75th anniversary of the establishment of diplomatic relations between China and Russia, with the support of the Russian Ministry of Culture, Nanshan Museum and The Russian Museum of Ethnography jointly held an exhibition *Cocoon of Traditions: Silk in the National Cultures of Eurasia*. Although many ethnic groups have their own traditional silk preferences, they do not reject the introduction of precious silk products and materials that meet their social and cultural needs, and finally formed a unique "Cocoon of Traditions" through integration and innovation.

The Russian Museum of Ethnography is one of the largest ethnographic museums in the world, with more than 500,000 exhibits. The first collection began in the mid-1890s. The Russian Museum of Ethnography displays tools, household items, folk costumes, utensils and ritual items, allowing people to understand the occupations, residences, festivals, rituals and religious beliefs of local residents.

The 283 pieces/sets of exhibits on display in this exhibition are mainly concentrated between the 19th and 20th centuries, including robes made of Chinese silk, samples of Central Asian handmade fabrics, Iranian silk carpets, "Kragai" headscarves from the Southeast Caucasus, and rich silk embroidered costumes of the Finno-Ugric people in the Volga region. Through these exquisite silk fabrics, the audience can feel the diversity of machine-woven and handmade silk products in the cultures of various ethnic groups in Eurasia.

"Civilizations are colorful because of exchanges and rich because of mutual learning." This exhibition is one of the series of exhibitions on civilizations along "the Belt and Road" held by Nanshan Museum. It is hoped that the exhibition will bring endless reverie to the audience about the land of Eurasia and fully demonstrate the phenomenon of exchanges and communication between civilizations along the Silk Road.

Qi Xin
Director of Nanshan Museum

俄罗斯民族博物馆馆长 致辞

尊敬的各位来宾:

欢迎各位参观俄罗斯民族博物馆的展览"传统之茧:欧亚大陆民族文化中的丝绸",本次展览旨在展示欧亚大陆中部的各民族丝绸织物在生产和流传过程中的特征。丰富多样的藏品向我们诉说着除中国以外仍在发展中的各个养蚕和丝织中心的情况,引领我们溯源欧亚大陆各地区之间的历史文化联系,让我们得以从大师的作品中汲取灵感,欣赏各国丝绸产品的美学和美感。

丝绸是人类历史上最古老、最昂贵、最负盛名的面料之一,作为一种人造产品和审美现象,它是力量和温柔、幸福和创造力的象征,彰显出文化传统的继承性与代代相传的重要性,吸引着人们的目光。制作和使用丝绸的传统已成为许多国家的标志性名片及其文化遗产的一部分。如今,无论是作为技术还是艺术,丝绸仍然广受欢迎。

俄罗斯民族博物馆的展览是中俄文化交流年项目的一部分。尤为重要的是,这场展览举办于中俄建交 75 周年之际,是两国合作、互利、建立深厚友谊的成果,是中俄人文纽带不断加强的生动体现。

感谢深圳市南山博物馆、湖北省博物馆、艺米(天津)文化传播有限公司的领导及工作人员、俄罗斯民族博物馆发展基金会以及本次展览的所有合作伙伴,是你们的通力合作促成了这次展览的成功举办。另外,也要特别感谢俄罗斯联邦文化部和中华人民共和国文化和旅游部为该项目提供的宝贵支持。

然而,展览的主要创造者始终是参观者。我真诚地希望此次展览能够在中俄两国文化交流活动中引发热烈反响,为参观者提供观察两国关系的新视角,触发未来的新合作。

尤利娅 · 库皮娜
俄罗斯民族博物馆馆长

Preface by the Director of the Russian Museum of Ethnography

Honorable guests,

I am glad to welcome you to the exhibition of the Russian Museum of Ethnography *Cocoon of Traditions: Silk in the National Culture of Eurasia*, dedicated to the peculiarities of the production and use of silk fabrics among the peoples of Central Eurasia. The richest collections of our museum tell about various centers of sericulture and silk weaving that arose outside of China, which are preserved and developed to this day. The exhibits of the exhibition allow you to trace the historical and cultural ties in the Eurasian space, get inspired by the work of craftsmen and enjoy the aesthetics and beauty of silk products of various peoples.

Silk, being one of the most ancient, valuable and prestigious materials in the history of mankind, always arouses our interest as a man-made product and an aesthetic phenomenon, as a symbol of strength and tenderness, well-being and creativity, as a symbol of the continuity of traditions and the importance of their transmission through generations. The traditions of making and using silk have become the calling card of many nations, part of their cultural heritage. They remain in demand today both as technologies and as art.

Our museum's exhibition is part of the Years of Culture program between China and Russia. What is particularly important is that this exhibition is held on the occasion of the 75th anniversary of the establishment of diplomatic relations between China and Russia, being the result of cooperation, mutual interest and friendship between our countries, a vivid manifestation of our growing cultural ties.

We are grateful for the creative cooperation to the management and staff of the Nanshan Museum and the Hubei Provincial Museum, Yimi Culture and Art Communication (Tianjin) Co., Ltd. and the Russian Museum of Ethnography Development Foundation, as well as to all partners who made this exhibition possible. The project received invaluable support from the Ministry of Culture of the Russian Federation and the Ministry of Culture and Tourism of the People's Republic of China.

However, the main creator of the exhibition is always the visitor. I sincerely hope that the exhibition will become a notable event in the series of cultural exchanges between China and Russia, that it will provide visitors with a new perspective on the relationship between the two countries and inspire future collaboration projects.

Dr. Yulia Kupina
Director of the Russian Museum of Ethnography

目录
Contents

前言
Foreword

　　众所周知，蚕丝业的发源地为古代中国，优质的中国丝绸逐渐征服了世界。丝绸的表现形式多种多样，但其主要特征是具有极高的标志性地位：丝绸制品卓越、繁荣和神圣的纯洁性。然而，对于许多民族来说，丝绸制品也是日常服饰的一部分。从事丝绸生产的民族至今仍将丝织传统视为其璀璨的文化遗产，将丝织品视为民族的象征。

　　本次展览主要反映了 19 至 20 世纪初，欧亚大陆中部民族文化中机织丝制品和手工丝制品相结合的产物多样性。同时，致力于展示丝绸文化——人类历史上最精美、最昂贵的纺织材料，其价值可以与黄金媲美。丝绸被视为中国和欧亚大陆主要历史和民族地区——西伯利亚、中亚、伊朗、高加索、东欧等地区的地方文化标志，此次的展品均为俄罗斯民族博物馆的馆藏，包括用中国丝绸制作的长袍、中亚手工织物样品、伊朗丝毯、东南高加索地区的"克拉加依"头巾等种类丰富的丝绸制品。这种全方位的展示使每个地区丝绸服饰的独特性更加鲜明，并证明了即使是相邻地区，丝织品的使用方法也可能不尽相同。在许多民族中，虽然存在各自的丝绸传统偏好，但它们并不排斥引进适应其社会文化需求的名贵丝绸用品和材料，并通过融合和创新最终形成了别具一格的"传统之茧"。

　　＊俄罗斯民族博物馆注：本书中出现的部分地区名称使用 19 世纪沙皇俄国时期的行政地区划分。20 世纪，有部分省份被废除、划分或合并。

As we all know, the birthplace of the silk industry is ancient China, high-quality Chinese silk gradually conquered the world. The silk has various manifestations, but its main feature is its extremely iconic status: the excellence, prosperity and sacred purity of silk products. However, for many peoples, silk products are also part of daily clothing. Peoples engaged in silk production still regard the silk weaving tradition as their brilliant cultural heritage and silk products as a symbol of the nation.

This exhibition mainly reflects the diversity of the products combining machine-woven silk products and handmade silk products in the national cultures of Central Eurasia from the 19th to early 20th century. At the same time, it is committed to showing the silk culture - the most exquisite and expensive textile material in human history, whose value can be compared with gold. Silk is a symbol of local culture in the major historical and ethnic regions of China and Eurasia, such as Siberia, Central Asia, Iran, the Caucasus, and Eastern Europe. The exhibits are from the collection of the Russian Museum of Ethnography and include a wide range of silk products such as robes made of Chinese silk, samples of handmade fabrics from Central Asian, Iranian silk carpets, "Kragai" headscarves in the southeastern Caucasus. This all-round display makes the uniqueness of silk clothing in each region more distinct, and proves that even in close neighbors, the use of silk fabrics may be different from each other. Among many ethnic groups, although they have their own traditional preferences for silk, they do not exclude the introduction of precious silk items and materials adapted to their socio-cultural needs, and through integration and innovation ultimately formed a unique "cocoon of tradition".

★ The Russian Museum of Ethnography Notes: Some of the names of the regions appearing in this book were used for the division of administrative regions during the period of Tsarist Russia in the 19th century. In the 20th century, some provinces were abolished, divided or merged.

丝绸的起源、传播与交流
The Origin, Spread, and Exchange of Silk

    我国是世界上最早驯养家蚕、缫丝织绸的国家。关于丝绸的起源，我国史籍中有不少神话传说，据《通鉴纲目》记载，黄帝妃嫘祖"始教民育蚕，治丝茧以供衣服，而天下无皴瘃之患，后世祀为先蚕"。耶鲁大学历史学教授芮乐伟·韩森在其著作《丝绸之路新史》中明确指出"中国人确实是世界上第一个制造出丝绸的民族"。

    考古发现表明，中国可能在裴李岗文化时期就已经出现丝蛋白。20 世纪 80 年代，河南荥阳青台村仰韶文化遗址考古发掘了一块浅绛色罗织物，距今已有 5600 多年的历史，是迄今世界上发现的最早的丝织物，也与传说中嫘祖教民养蚕的时期相近。

China is the earliest country in the world to domesticate silkworms and weave silk. There are many myths and legends about the origin of silk in Chinese historical records. According to the *Outline of the Comprehensive Mirror*, Empress Leizu of the Yellow Emperor "first taught people to raise silkworms, treat silk cocoons for clothing, and no one in the world has frostbite anymore. Later generations worshipped silkworms as the first". Yale University history professor Valerie Hansen explicitly stated in her book *The Silk Road : A New History* that "the Chinese were indeed the first people in the world to produce silk".

Archaeological discoveries suggest that silk protein may have appeared in China during the Peiligang Culture period. In the 1980s, a light reddish silk fabric was excavated at the Yangshao Cultural Site in Qingtai Village, Xingyang, Henan Province. It has a history of over 5600 years and is the earliest silk fabric discovered in the world. It is also similar to the time when the legendary Leizu taught people to raise silkworms.

罗织物，约公元前 3650 年出土于河南荥阳青台村
Luowen fabric, unearthed in Qingtai Village, Xingyang, Henan, circa 3650 BCE

西方人将蚕称为"蚕儿"或"赛儿"，把养蚕的国家称为"赛里斯"，养蚕的人称为"赛里斯人"，"丝国"因此成为中国的代称。公元前5世纪至公元前3世纪，中国丝绸通过草原丝绸之路向外传播，其中重要的媒介就是沿阿尔泰山脉活动的斯基泰人。这一通道东起蒙古高原，翻越阿尔泰山，经准噶尔盆地到哈萨克丘陵，横贯东西。汉武帝时期，张骞两次出使西域联系西域各国共同抗击匈奴，张骞出使的线路被称为"沙漠绿洲丝绸之路"。中国的蚕丝和丝绸沿着这条丝绸之路源源不断地传到中国西北，再由这里输往中亚、西亚并到达罗马帝国境内。中国的丝绸生产技术也在这一时期传到中亚。

　　丝绸是中西方文化交流的重要载体，丝绸由东到西传播的过程也在不断"胡化"。首先是丝绸生产技术的当地化；其次是中国传统织物受到萨珊波斯、索格狄亚纳、印度等外来文化的影响，出现极具异域风格的纹样。

Westerners refer to silkworms as "Can'er" or "Sai'er", refer to the country where silkworms are raised as "Sairis", and refer to the people who raise silkworms as "Sairis people". "Silk Country" has become synonymous with China. From the 5th century BCE to the 3rd century BCE, Chinese silk spread outward through the grassland Silk Road, and the important medium was the Scythians who moved along the Altai Mountains. This passage starts from the Mongolian Plateau in the east, crosses the Altay Mount Taishan Mountains, passes through the Junggar Basin, and reaches the Kazakh hills, traversing from east to west. During the reign of Emperor Wu of Han, Zhang Qian twice sent envoys to the Western Regions to connect with various countries in the Western Regions to jointly resist the Xiongnu. The route of Zhang Qian's mission was known as the Desert Oasis Silk Road. Chinese silkworms and silk were continuously transmitted along this Silk Road to the northwest of China, and then transported to Central Asia and West Asia, reaching the territory of the Roman Empire. China's silk production technology also spread to Central Asia during this period.

Silk is an important carrier of cultural exchange between China and the West, and the process of silk spreading from East to West is also constantly being "foreignized". Firstly, its manifestation is the localization of silk production technology; secondly, traditional Chinese fabrics have been influenced by foreign cultures such as Sassanid Persia, Sogdians and India, resulting in highly exotic patterns.

新疆丹丹乌里克寺院遗址出土的"蚕种西传"木板画
"Silkworm Seed Transmission to the West" Wooden Board Painting, unearthed at the Ulik Temple Site in Dandan, Xinjiang

19 世纪前，欧亚大陆的丝绸文化传播
The Spread of Silk Culture in Eurasia before the 19th Century

在古代世界和中世纪早期，丝绸文化主要沿北纬 40 度在欧亚大陆传播。在索格狄亚纳、古伊朗、历史上的亚美尼亚和拜占庭形成了大片的养蚕区。中世纪晚期，奥斯曼帝国的苏丹宫廷继承了拜占庭传统。17 世纪，在亚美尼亚商人的斡旋下，萨非王朝统治下的伊朗在世界丝绸贸易中占有一席之地。

During ancient history and the Early Middle Ages, the silk culture spread across Eurasia, mainly along the 40th parallel. Large sericulture zones sprung up in Sogdia, ancient Iran, historical Armenia, and Byzantium. In the Late Middle Ages, the Byzantine traditions were inherited by the Sultan's court of the Ottoman Empire. Safavid Iran in the 17th century, in cooperation with Armenian merchants, took the leading positions in the world silk trade.

地毯，萨非王朝，16 世纪，丝绸，
大都会艺术博物馆
Carpet, Safavid, 16th century, silk,
The Metropolitan Museum of Art

祈祷地毯，奥斯曼，1575 ~ 1590 年，丝（经线和纬线）、羊毛（绒毛）、棉（绒毛），大都会艺术博物馆
Prayer Carpet, Ottoman, 1575-1590, silk (warp and weft), wool (pile), cotton (pile), The Metropolitan Museum of Art

到彼得大帝统治下的俄罗斯帝国时期，亚美尼亚企业家出资在阿斯特拉罕建立了第一批丝绸种植园和丝绸纺织厂。18 世纪中叶，莫斯科也成为丝织中心。在帝国经济利益区的扩张过程中，中亚、南高加索、北黑海沿岸的丝绸地区仍然存在手工丝织业。

By the time of the Russian Empire of Peter the Great, the first silk plantations and silk-weaving manufactories were founded in Astrakhan, sponsored by Armenian entrepreneurs. In the mid-18th century, Moscow also became a sericulture center. As the empire was getting bigger, the silk-growing areas of Central Asia, the South Caucasus, and the Northern Black Sea region, with their artisanal silk-weaving workshops, got into the zone of its economic interests.

从事生丝贸易的亚美尼亚商人约翰·魏格尔，17 世纪
Johann C. Weigel, an Armenian merchant engaged in raw silk, 17th century

I

俄罗斯各地区的
丝绸文化

Silk Culture of
the Russian Regions

年轻女人
俄罗斯人，科斯特罗马省
1907年
Young woman
Russians. Kostroma Province
1907

穿着节日盛装的妇女
俄罗斯人，下诺夫哥罗德省，下乌金斯克地区
1867年
Women in festive costumes
Russians. Nizhniy Novgorod Province, Nizhneudinsk Uyezd
1867

俄罗斯北部和中部省份居民文化中的丝绸
Silk in the Culture of the Russian Population of Russia's Northern and Central Provinces

19 至 20 世纪初，俄罗斯工厂生产的各种丝织品主要出现在俄罗斯北部和中部省份、乌拉尔和西伯利亚地区。人们用丝绸缝制衬衫、"萨拉凡"（俄罗斯传统服装，一种无袖宽松长裙）、有肩带或袖子的短款胸衣以及上衣——"舒佳衣"（俄罗斯传统服装，一种对襟短袖上衣，衣尾有褶皱）、斗篷款披肩、短款皮草、男式长袍，也用于做帽饰的底布和裘皮大衣的内衬面。

From the 19th century to the early 20th century, a variety of silk fabrics, usually machine-made in the northern and central provinces of Russia, the Urals, and Siberia. Silk was the fabric of choice for shirts, Sarafans (Russian traditional dress, a sleeveless, loose-fitting dress), short breast wear with straps or sleeves, and outwear, like Shugai (Russian traditional dress, a short-sleeved blouse with lapels and pleats at the end of the garment), pelerines, fur coats, and kaftans, they were also used to cover headdress bases and undersides of fur coats.

女士护胸（"舒佳衣"）

俄罗斯人，东欧，阿尔汉格尔斯克州 | 19 世纪 60 ~ 90 年代 | 厚丝，花布，丝绸，金属

Shugai – woman sleeveless jacket

Russians. Eastern Europe, Arkhangelsk Province | 1860-1890s | Heavy silk, printed calico, silk, metal

披肩是年轻女性的节日盛装中最亮眼、最引人注目的服饰之一。这种披肩尺寸较大，挂在头饰上，两端自然垂下，几乎垂到地面，通常能够完全遮住女性的身形。这类披肩在民间传统中有很多别称，比如头巾、围巾、披巾、罩布、"卡纳瓦特"纱巾、面纱等。其中，"卡纳瓦特"纱巾非常流行，这是一种红色调的条纹或棋盘格织物，图案为钻石、玫瑰花结，或用金属丝线绣制的花束。即便它们是在俄罗斯工厂生产的，但这种披肩的灵感来自于东方织物，其流传范围十分广泛。

One of the most vibrant and prominent elements of a young woman's festive attire was a shawl that was fixed onto a headdress. Shawls were of considerable sizes, they would hug a headdress so that their ends dangled down freely, almost reaching the floor, often completely enveloping a woman's silhouette. The folk tradition knows such items as headscarves, scarves, shawls, covers, qanawats (kanawats), or veils. Qanawats were especially popular, they presented striped or checkered fabrics of red shades with a pattern in the form of diamonds, rosettes, or bouquets created using a metal thread. Though made at Russian factories, they were inspired by oriental fabrics, the geography of qanawats' popularity was quite vast.

披肩

俄罗斯人，东欧，科斯特罗马省 | 19 世纪下半叶 | 丝绸，丝线，金线

Kerchief

Russians. Eastern Europe, Kostroma Province | Second half of the 19th century | Silk, silk thread, gold thread

披肩

俄罗斯人，东欧 | 19 世纪下半叶 | 丝绸，金属混合纱，黄色合金

Kerchief

Russians. Eastern Europe | Second half of the 19th century | Silk, metallic combination yarn, yellow alloy

披肩

俄罗斯人，东欧 | 19 世纪下半叶 | 丝绸，金属混合纱，黄色合金

Kerchief

Russians. Eastern Europe | Second half of the 19th century | Silk, metallic combination yarn, yellow alloy

俄罗斯女性是否使用进口纺织品，尤其是丝绸，取决于家庭的物质财富水平和穿着场合的仪式化程度。在大型节日或婚礼上，年轻女性的服装几乎都是用丝绸面料制成的——真丝、锦缎、"柳斯特林"（一种有光泽的丝织物）、塔夫绸、绸缎、天鹅绒、织锦、半织锦等，并用丝带或金银绦带花边做装饰。

此外，用"萨拉凡"和"舒佳衣"组成的套装，也是俄罗斯女性在节日中喜欢穿戴的服饰。这种锦缎还被用来制作上衣、帽饰、手套、手提包和毛皮大衣的表面。

Whether the Russian women incorporated imported fabrics, especially silk, or not, was determined by the material well-being of a family and the degree of ceremonialism of the situation it was worn to. Quite often young married women had their festive attires and wedding sets almost completely made of silk fabrics, like silk, damask, lustrine, taffeta, satin, velvet, brocade, brocatelle, etc., decorated with silk or passementerie ribbons.

In addition, a popular variant of a festive girl or woman's costume was a Sarafan and Shugai set. The same damask was used for outerwear, hats, gloves, handbags, and to cover fur coats.

017

年轻女士的节日服装
俄罗斯人
19 世纪中叶
Young woman's festive costume
Russians
Mid-19th century

女士头饰

俄罗斯人，东欧
19 世纪下半叶
丝绸，亚麻布，金属混合纱，玻璃珠

Woman headwear

Russians. Eastern Europe
Second half of the 19th century
Silk, linen, metallic combination yarn, glass beads

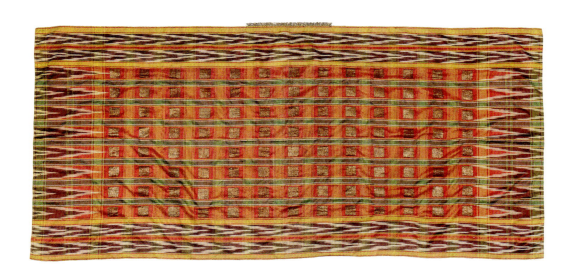

头纱

俄罗斯人，东欧 ｜ 19 世纪 ｜ 丝绸，金属混合纱

Veil

Russians. Eastern Europe ｜ 19th century ｜ Silk, metallic combination yarn

女士外套（"舒佳衣"）

俄罗斯人，东欧 ｜ 19 世纪下半叶 ｜ 厚丝，丝绸，手工织染布，白色合金

Shugai – woman outerwear

Russians. Eastern Europe ｜ Second half of the 19th century ｜ Heavy silk, silk, dyed thick linen of plant fiber, white alloy

女士无袖长裙（"萨拉凡"）

俄罗斯人，阿尔罕格尔斯克省 | 19 世纪 70 ~ 90 年代 | 丝绸，厚丝，硬衬布

Sarafan – Woman sleeveless dress

Russians. Arkhangelsk Province | 1870–1890s | Silk, heavy silk, buckram

年轻女士的节日服装
俄罗斯人
19 世纪初 ~ 19 世纪中叶
Young women's festive costumes
Russians
Early 19th - mid 19th century

女士短外套

俄罗斯人 | 19 世纪 20 ～ 30 年代 | 锦缎，棉织物，黄色金属，印花布，金线

Woman outerwear

Russians | 1820–1830s | Brocade, cotton, yellow metal, calico, gold thread

女士无袖短上衣

俄罗斯人 | 1900 ~ 1925 年 | 锦缎，棉织物，金线

Woman sleeveless short top

Russians | 1900-1925 | Brocade, cotton, gold thread

女士头饰

俄罗斯人
19 世纪中叶
玻璃，麻布，丝绸，红布

Woman headwear

Russians
Mid-19th century
Glass, burlap, silk, red cloth

女士头饰（头巾）

俄罗斯人 | 19 世纪下半叶 | 棉织物，金线

Woman headwear (veil)

Russians | Second half of the 19th century | Cotton, gold thread

哥萨克民族服饰中的丝绸
Silk in Cossack National Costume

在俄罗斯南部，丝绸织物通常出现在顿河哥萨克和捷列克哥萨克妇女的服装中，这类服饰与北高加索人的服饰相似，有袖长至手腕的丝绸衬衫和收腰连衣裙，人们用丝绸、锦缎、绸缎、塔夫绸和锦花绸缝制节日服装，有时也用于日常服饰。顿河哥萨克女性的服饰称作"库别列克"，这是一种对襟服装，腰部有剪裁，上衣紧身，胸前有排列紧密的扣子，裙摆宽大。她们还会在脑后盘好的头发上戴一顶丝纹尖顶帽来搭配服饰。捷列克哥萨克女性在衬衫外也会穿一件类似"库别列克"的外衣，但衣服领口较宽。为了遮住头部，她们使用素色和有花纹的丝绸披肩，并将其覆盖在双角帽上。

对于顿河和北高加索的哥萨克女性来说，身着华丽的丝绸和锦缎面的狐狸毛、松鼠毛、兔毛大衣是一件值得骄傲的事，上面覆盖着丝绸或锦缎织物。这种大衣底部加宽，穿戴时可以用腰带包裹或系住，因款式特别被称为"顿河皮草"。

前高加索草原的哥萨克人通常用本地区生产的丝绸制作衣服，尽管欧洲和亚洲的丝绸编织面料在该地区并不罕见。在18到19世纪期间，捷列克哥萨克人积极从事养蚕业以供家用或出售。哥萨克妇女将自产的丝线编织成腰带、装饰带、花边和线绳，用来装饰服装。

On the territory of the southern part of Russia, silk fabrics were most often found in the costumes of the Don and Terek Cossack women. Their clothes, outwardly resembling outfits of the North Caucasus peoples, included a long silk shirt with sleeves extending to the wrists, and an outwear slim fit dress, festive and sometimes everyday variations of which were made of silk, damask, satin, taffeta, and others. The dress of the Don Cossack women was called kubelek, it was a kind of an unfastened garment, often waist-cut, with a tight-fitting bodice and a lot of chest clasps and a wide skirt. They would also wear a silk-patterned peaked cap over their coiled hair at the back of their heads to match their costumes. The Terek Cossack women wore beshmets over their shirts, similar to kubeleks, but with a wider neckline. To cover their heads, they used plain and patterned silk headscarves worn over horned kichkas.

The object of pride for the Cossack women at Don and the North Caucasus was an elegant fur coat made of fox, squirrel and hare furs, covered with silk or damask fabric. The fur coat got wider to the bottom and when worn was supposed to be wrapped or tied with a kushak (sash belt). These fur coats were called "Don coats" for their specific silhouette.

The Cossacks of the pre-Caucasian steppes, as a rule, made their clothes from silk of domestic production, although European and Asian silk-weaving fabrics were not rare in the area. During the 18th and 19th centuries, the Terek Cossacks were actively engaging in sericulture for home use or sale. The Cossack women utilized home-produced silk thread to weave belts, passementerie and galloons, cords and webbing that were used as garment adornments.

哥萨克和顿河哥萨克
Cossacks and Don Cossacks

哥萨克主要是东斯拉夫东正教民族，起源于乌克兰东部和俄罗斯南部的庞迪–里海草原。历史上，他们是半游牧、半军事化的民族，虽然名义上受当时东欧各国的宗主国管辖，但以服兵役作为交换条件，他们曾享有很大程度的自治权。哥萨克团体按照军事线路组织起来。他们主要分布在第聂伯河、顿河、捷列克河和乌拉尔河流域人烟稀少的地区。

顿河哥萨克则是一个半游牧民族，生活在俄罗斯西南部顿河中下游地区。他们以独特的军事文化、精湛的马术和残酷的武士习俗而闻名。

The Cossacks are a predominantly East Slavic Orthodox Christian people originating in the Pontic-Caspian steppe of eastern Ukraine and southern Russia. Historically, they were a semi-nomadic and semi-militarized people, who, while under the nominal suzerainty of various Eastern European states at the time, were allowed a great degree of self-governance in exchange for military service. Cossack groups were organized along military lines. They inhabited sparsely populated areas in the Dnieper, Don, Terek, and Ural river basins.

Don Cossacks are a semi-nomadic people, which settled along the middle and lower Don river in southwestern Russia. They were famous as bearers of unique military culture, superb horsemanship and brutal warrior customs.

027

顿河哥萨克地区的节日服装
俄罗斯人
19 世纪末 ~ 20 世纪初
Festive costumes of the Don Cossack Host Province
Russians
Late 19th - early 20th century

已婚妇女"顿河长裙"

俄罗斯人，顿河哥萨克地区 ｜ 19 世纪 70 ~ 90 年代 ｜ 棉织物，丝绸，羊毛，铜合金，玻璃

Married woman dress

Russians. Don Cossack Host Province ｜ 1870-1890s ｜ Cotton, silk, wool, copper alloy, glass

腰带
俄罗斯人，顿河哥萨克地区
19 世纪中叶
天鹅绒，丝绸，半丝织物，玻璃，铜

Belt
Russians. Don Cossack Host Province
Mid-19th century
Velvet, silk, semi-silk, glass, copper

029

头巾
俄罗斯人，东欧，科斯特罗马省，科罗格里夫地区 | 20 世纪初 | 丝绸

Head kerchief
Russians. Eastern Europe, Kostroma oblast, Kologrivsky district | Early 20th century | Silk

俄罗斯南部地区服饰中的丝绸
Silk in the Costume of the Southern Regions of Russia

19 世纪中叶起，俄罗斯南部地区的农民服饰开始频繁使用丝线和丝织品做装饰。直到 20 世纪，俄罗斯南部地区女性穿着的波涅瓦裙套装都是用自家生产的布料制作的。人们不仅用丝线、丝带、花边和剪裁的布料装饰服装，还将丝绸缎带广泛用于波涅瓦裙、围裙、胸衣、头饰等服饰。因此，女孩的头巾都配有丝绸装饰，部分地区的女孩头饰像对角折叠的丝绸手帕一样简洁，丝绸头饰也是农村年轻女性节日服饰中的重要组成部分。

Since the mid-19th century, silk thread and fabrics were actively used to decorate peasant clothes in the southern regions of Russia. Until the 20th century, items of the South Russian women's poneva skirt costume sets were made of home-produced fabrics. People not only decorate their garments with silk thread, ribbons, webbing and fabric cuts, silk ribbons are also widely used to decorate garments, like ponevas, aprons and nagrudniks, as well as headdresses. Thus, silk adorned girl's headbands, in some regions, a girl's headdress was as simple as a silk headscarf folded diagonally, silk headscarves were also an essential part of the festive costume set of a young peasant woman.

19 世纪俄罗斯女性服饰
Russian Women's Costumes of the 19th Century

科科什尼克斯（一种俄罗斯女帽）是俄罗斯服装中装饰内容最丰富的。它通常由金属线编织的锦缎或天鹅绒制成。富裕的农民阶层用珍珠和宝石装饰他们的帽饰，其装饰元素带有鲜明的地域色彩。北方的女帽用当地盛产的淡水珍珠作装饰，而南方则更流行鹅绒和羊毛刺绣装饰。少女佩戴头饰时会露出她们的头发，并同时佩戴用织物或金属制成的前额遮盖物，而已婚妇女必须完全遮住头发。右图中展示的头饰使用半宝石和彩色箔片作为装饰。

Kokoshniks (a Russian women's headdress) had the greatest abundance of ornamentation of any type of garment in Russia. They were most often made of damask woven with gilt metallic thread or velvet. The wealthy peasant class decorated their kokoshniks with pearls and gemstones. Their decorative elements were representative of the regions in which they were made. Those from the North were embellished with the river pearls that were plentiful in that area while goose down and woolen embroideries were more popular in the South. The headdresses worn by maidens exposed their hair and were often accompanied by a venchik, a forehead covering made of fabric or metal. Married women were required to cover their hair entirely. The headdress shown on the right is decorated with semi-precious stones and coloured foil.

19 世纪俄罗斯女性服饰 大都会艺术博物馆藏
Russian women's costumes of the 19th century
(Collection of the Metropolitan Museum of Art)

还有另外一种形制的俄罗斯女帽也是俄罗斯已婚妇女服饰中的重要组成部分（如右图所示）。这种头饰在俄罗斯中部也有发现。角状装饰的作用是保护妇女免受邪灵侵扰，同时也象征着妇女的青春和生育能力。

There is also another form of Russian women's headdresses that is an important part of Russian married women's dress (shown in the photo on the right). This headdress was also found in central Russia. The horns were meant to protect the woman from evil spirits and also symbolized her youth and ability to bear children.

031

19 世纪俄罗斯女性帽饰 俄罗斯民族博物馆藏
Russian women's headdresses of the 19th century
(Collection of the Russian Museum of Ethnography)

头纱

俄罗斯人，东欧
19 世纪上半叶
丝绸，金属混合纱

Veil

Russians. Eastern Europe
First half of the 19th century
Silk, metallic combination yarn

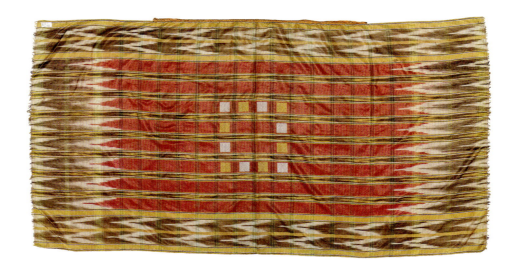

头纱

俄罗斯人，东欧，坦波夫省，利佩茨克地区 ｜ 19 世纪 ｜ 丝绸，金线

Veil

Russians. Eastern Europe, Tambov Province, Lipetsk Uyezd ｜ 19th century ｜ Silk, gold thread

头巾

俄罗斯人，东欧，科斯特罗马省 ｜ 19 世纪下半叶 ｜ 丝绸，丝线，金属混合纱

Kerchief

Russians. Eastern Europe, Kostroma Province ｜ Second half of the 19th century ｜ Silk, silk thread, metallic combination yarn

034

头巾

俄罗斯人，东欧，伏尔加格勒省，索尔维切戈特市 | 19 世纪下半叶 | 丝绸，金属混合纱

Kerchief

Russians. Eastern Europe, Vologda Province, Solvichegodsk Uyezd | Second half of the 19th century | Silk, metallic combination yarn

年轻女士头巾

俄罗斯人，东欧
19 世纪
亚麻布，丝线

Young woman kerchief

Russians. Eastern Europe
19th century
Linen, silk thread

年轻女士头巾

俄罗斯人，东欧 ｜ 18 世纪末 ｜ 塔夫绸，金属混合纱，丝线

Young woman kerchief

Russians. Eastern Europe ｜ Late 18th century ｜ Taffeta (silk), metallic combination yarn, silk thread

女士腰部服装局部

俄罗斯人，东欧，沃罗涅日省，比柳琴地区

19 世纪下半叶

羊毛，生丝

Part of the woman bottomwear

Russians. Eastern Europe, Voronezh Province,

Biryuchensky Uyezd

Second half of the 19th century

Wool, raw-silk

腰带

俄罗斯人 | 19 世纪 | 丝绸

Belt

Russians | 19th century | Silk

037

腰带

俄罗斯人，东欧，奥罗涅茨省 | 19 世纪 | 丝绸

Belt

Russians. Eastern Europe, Olonets Province | 19th century | Silk

女士头饰

俄罗斯人，东欧，梁赞省，拉恩堡地区
19 世纪 70 ~ 90 年代
亚麻布，棉织物，羊毛，丝线

Woman headwear

Russians. Eastern Europe, Ryazan Province,
Rarnborough area
1870-1890s
Linen, cotton, wool, silk thread

女士头饰

俄罗斯人，东欧，坦波夫省 | 1875 ~ 1900 年 | 亚麻布，红印花布，棉织物，丝绸，玻璃，金属

Woman headwear

Russians. Eastern Europe, Tambov Province | 1875-1900 | Linen, red calico, cotton, silk, glass, metal

女孩头饰局部

俄罗斯人，东欧，坦波夫省

19 世纪 70 ～ 90 年代

棉织物，丝绸，丝线，木，黄色金属

Part of a young woman headwear

Russians. Eastern Europe, Tambov Province

1870-1890s

Cotton, silk, silk thread, wood, yellow metal

丝带

俄罗斯人，东欧，坦波夫省，利佩茨克地区

19 世纪末

丝，棉织物，丝线，金属线

Ribbon

Russians. Eastern Europe, Tambov Province, Lipetsk Uyezd

Late 19th century

Silk, cotton, silk thread, metal thread

俄罗斯城乡地区中的丝绸配饰
Silk Accessories in Russian Urban and Rural Areas

　　俄罗斯生产的丝绸面料以朴素和价格低廉为特点，其主要面向中产的城乡居民。人们可以在集市、市场及商铺中购买到剪裁过的面料，也可以直接购买一些现成的丝绸产品，如披肩、罩布、腰带、手套、钱包、手提包、丝带、花边、线绳和流苏穗边等。

　　除了布料和成品外，女性往往会购买丝线用于刺绣。丝绸刺绣被认为是男女头饰、鞋子、配饰、室内物品和教堂用具流行的装饰。17 至 18 世纪早期的俄罗斯上层社会也十分盛行丝绸刺绣传统。大多数情况下，仪式中通常使用有丝线装饰的象征性物品，包括希林卡披肩。丝绸刺绣还用来装饰室内纺织品，比如毛巾、桌布、床围等，它们在民间传说中被认为具有神奇的功能。直到 19 至 20 世纪初，这类物品在民间仍发挥着重要作用，农民和部分地区的城镇居民将其作为先辈留下的遗产用心保护并世代相传。

Silk fabrics produced in Russia had a few distinctive features, like their simplicity and cheapness, they were intended for townspeople and peasants of moderate means. Fabrics were available for purchase at fairs, markets, and shops sold by cuts, then they were used to make clothes and accessories, it was also possible to buy some ready-made silk products, like shawls and covers, belts, gloves, purses and handbags, ribbons and webbing, cords and tassels.

Along with silk fabrics and goods, silk thread were sold that women purchased for needlework. Silk embroidery was considered a popular garment decoration for both men's and women's headdresses, shoes, accessories, as well as interior items and church utensils. The silk embroidery tradition was quite familiar to the upper classes of the Russian society of the early 17th-18th centuries, too. Most often, iconic objects decorated with silk thread, including shirinka shawls, were commonly used in rituals. Silk embroidery was also common to textile interior items, such as towels, tablecloths and bed skirts, which, in folklore, also performed magical functions. From the 19th to early 20th century, all these items remained significant mainly as folk objects. Peasants and sometimes townspeople tended after them carefully, they inherited them from their elders and passed them on to their children and grandchildren.

19 世纪末、20 世纪初的俄罗斯城乡移民问题
Problem of Russian Peasant Migration in the Late 19th and Early 20th Centuries

在俄国农奴制改革后的几十年里，随着乡村就业机会的减少和对现金的追求，越来越多的农民离开家乡，前往城市和工业中心。在第一次世界大战之前，移民中的绝大多数都是男性，女性依然是乡村生活中的中坚力量。尽管如此，在 19 世纪即将结束之际，促使男性离开村庄的经济环境也导致女性离开村庄。像她们的丈夫、父亲和兄弟一样，移民女性通常选择城市作为目的地。

In the decades following the emancipation of the serfs, increasing numbers of peasants left their native villages for cities and industrial centers, in response to a growing need for cash and declining opportunities to earn it at home. At least until World War I, the vast majority of these migrants were men, women were the more stable element in the village. Nevertheless, as the nineteenth century drew to a close the economic circumstances that prompted peasant men to leave villages increasingly caused women to leave as well. Like their husbands, fathers and brothers, migrant women often chose urban destinations.

服饰中的刺绣装饰
Embroidery in Costumes

刺绣的使用方式与文身大致相同，都是用象征权力的图案来保护穿着者。正因如此，刺绣图案通常装饰在服装边缘，如颈部与腕部。刺绣还经常被装饰在象征力量的上臂肌肉上。

Embroidery is used in much the same way as tattooing to protect the wearer with symbols of power. It is for this reason that embroidery motifs are so commonly placed around the edges of garments, such as the neck and wrists. Embroidery is also often added over the upper arm muscles, symbolizing strength.

手套

俄罗斯人，东欧
19 世纪下半叶
厚丝，印花棉布，丝绸，金线，丝线

Gloves

Russians. Eastern Europe
Second half of the 19th century
Heavy silk, printed calico, silk, gold thread, silk thread

女士手套

俄罗斯人 | 19 世纪下半叶 | 丝绸，丝线，金属线

Woman Gloves

Russians | Second half of the 19th century | Silk, silk thread, metallic cord

衣袋

俄罗斯人，东欧，伏尔加格省

19 世纪 60 ～ 90 年代

亚麻布，丝绸，丝线

Pocket

Russians. Eastern Europe, Vologda Province

1860–1890s

Linen, silk, silk thread

毛巾

俄罗斯人，东欧 | 19 世纪下半叶 | 亚麻布，金属混合纱，丝线

Towel

Russians. Eastern Europe | Second half of the 19th century | Linen, metallic combination yarn, silk thread

床单边缘

俄罗斯人，俄罗斯（欧洲部分）

18 世纪

厚丝，印花布，丝线，金属混合纱

Bed skirt

Russians. European Russia

18th century

Heavy silk, calico, silk thread,
metallic combination yarn

床单边缘

俄罗斯人，俄罗斯（欧洲部分）

19 世纪下半叶

亚麻布，丝带，捻丝，亚麻线

Bed skirt

Russians. European Russia

Second half of the 19th century

Linen, ribbon, silk twisted thread,
linen thread

045

2

南西伯利亚和远东地区
民族传统文化中的丝绸

Silk in the Culture of
the Peoples of Southern Siberia
and the Far East

刺绣师在工作
那乃人，远东边疆地区，
尼古拉耶夫—阿穆尔河畔
1927年
Embroideress during a work
Nanai. Far Eastern Krai,
Nikolaevsky-na-Amure Okrug
1927

毡房里的一家人
阿尔泰—铁列族人，托木斯克州，比斯克区
1903年
The family near felt yurt
Altaians - Telengits. Tomsk Province, Biysk District
1903

突厥-蒙古文化各民族的丝绸
Silk in the Peoples of the Turko-Mongol Cultures

　　长达几世纪以来，突厥-蒙古文化的各民族（蒙古人、布里亚特人、卡尔梅克人、图瓦人、阿尔泰基齐人、铁列人、哈卡斯人）一直使用中国丝绸制作节日盛装、名贵服饰、马匹及室内装饰。丝绸既用于为社会精英阶层缝制服装，也用于覆盖羊皮大衣。得益于贸易发展，丝织品和丝绸生产都融入游牧民族的文化中：多色锁针和平针刺绣已成为哈卡斯民族文化的象征之一；花卉成为主要装饰图案；南西伯利亚原住民服饰的特点是用彩色镶边和线绳进行装饰。

　　卡尔梅克人将西部蒙古瓦剌人的文化带到了伏尔加河下游的大草原，随着时间的推移，他们的服饰明显受到北高加索和俄罗斯传统的影响。与大多数游牧民族一样，卡尔梅克人使用进口纺织物（通常是丝绸）制作衣服。一个独特的细节是已婚的卡尔梅克女性辫子上的黑色布罩通常是用绸缎制成的。

For centuries it was peculiar for peoples of the Turko-Mongol cultures (Mongolians, Buryats, Kalmyks, Tuvans, Altai-Kiji, Telengits, Khakas) to use Chinese silk for making festive and prestige costume pieces, ceremonial decorative trappings of a horse as well as interior decoration. Silk fabrics were meant for the elite's garments and covering sheepskin fur coats. Due to trade expansion both silk fabrics and silk production integrated into nomadic culture: Polychromatic chain stitch and plain stitch embroidery became one of the national cultural symbols of the Khakas; floral became the main decorative motif; it was peculiar for the indigenous peoples of Southern Siberia to decorate their costumes with colored piping and cording.

The Kalmyks brought the culture of the western Mongols Oirats with them to the Steppes of the Low Volga region. Later in time, their costumes absorb the Northern caucasus and Russian traditions. Like most of the nomadic peoples, the Kalmyks made their clothes of imported fabrics, very often of silk. The peculiar detail of a married Kalmyks woman's costume is black plait cases usually made of satin silk.

布里亚特人
Buryats

　　布里亚特人是居住在西伯利亚东南部的蒙古族人。他们是西伯利亚最大的两个土著民族之一（另一支是雅库特人）。目前，大部分布里亚特人生活在布里亚特共和国。在布里亚特西部和东部的乌斯特-奥尔达布里亚特自治区（伊尔库茨克州）、阿金-布里亚特自治区（后贝加尔边疆区），以及蒙古国东北部和中国内蒙古也居住着较小规模的布里亚特人，他们构成了蒙古人的主要北方支系。

　　布里亚特人与其他蒙古人共享许多习俗，包括游牧和搭建蒙古包的习俗。如今，大多数布里亚特人居住在布里亚特共和国首府乌兰乌德及其周边地区，但仍有许多人在农村过着更为传统的生活。他们讲一种被称作布里亚特语的中部蒙古语。

The Buryats are a Mongolic ethnic group native to southeastern Siberia. They are one of the two largest indigenous groups in Siberia, the other being the Yakuts. The majority of the Buryats today live in the Republic of Buryatia. Smaller groups of Buryats also inhabit Ust-Orda Buryat Okrug (Irkutsk Oblast) and the Agin-Buryat Okrug (Zabaykalsky Krai) which are to the west and east of Buryatia as well as northeastern Mongolia and Inner Mongolia, China. They constitute the main northern subgroup of the Mongols.

Buryats share many customs with other Mongols, including nomadic herding, and erecting gers for shelter. Today the majority of Buryats live in and around Ulan-Ude, the capital of the Buryat Republic, although many still follow a more traditional lifestyle in the countryside. They speak a central Mongolic language called Buryat.

图瓦人
Tuvans

　　图瓦人是居住在俄罗斯、蒙古国和中国的西伯利亚突厥族群。图瓦语是一种西伯利亚突厥语。在蒙古国，他们被视为乌梁海民族之一。

　　图瓦人历史上一直是以畜牧为主的游牧民族，数千年来一直放牧山羊、绵羊、骆驼、驯鹿、牛和牦牛。他们居住在用毡或毡垫覆盖的蒙古包里，毡垫上铺着桦树皮或兽皮。图瓦人主要分布在以下这些地区，即托珠、萨尔恰克、奥尤纳尔、克姆奇克、哈苏特、沙利克、尼巴兹、达万、乔杜以及比齐。

The Tuvans are a Turkic ethnic group indigenous to Siberia who live in Russia (Tuva), Mongolia, and China. They speak Tuvan, a Siberian Turkic language. In Mongolia, they are regarded as one of the Uriankhai people groups.

Tuvans have historically been livestock-herding nomads, tending to herds of goats, sheep, camels, reindeer, cattle, and yaks for the past thousands of years. They have traditionally lived in yurts covered by felt or chums, layered with birch bark or hide that they relocate seasonally as they move to newer pastures. Traditionally, the Tuvans were found in these regions, namely the Tozhu, Salchak, Oyunnar, Khemchik, Khaasuut, Shalyk, Nibazy, Daavan, Choodu and Beezi.

卡尔梅克人
Kalmyks

　　卡尔梅克人是欧洲唯一的讲蒙古语的民族，居住在欧洲平原的最东部。这片位于伏尔加河下游以西、里海西北岸盆地的干燥草原地区，是最适合游牧的牧场。卡尔梅克人的祖先是讲卫拉特语的游牧民族，他们曾三次从蒙古国西部迁徙到东欧。他们是欧洲境内唯一信仰佛教的民族。

Kalmyks are the only Mongolic-speaking people living in Europe, residing in the easternmost part of the European Plain. This dry steppe area, west of the lower Volga River, basin on the Northwest shore of the Caspian Sea was the most suitable land for nomadic pastures. The ancestors of Kalmyks were nomads who spoke the Werat language, and migrated three times from Western Mongolia to Eastern Europe. They are the only traditionally Buddhist people who are located within Europe.

欧亚草原"突厥—蒙古族系"牧民的四种游牧模式
Four Nomadic Modes of the "Turko-Mongol" Herders of the Eurasian Steppes

　　苏联学者谢维扬·魏因施泰因指出，欧亚草原的"突厥-蒙古族系"牧民有四种游牧模式。第一种是"平原—山区—平原"型：冬季住平原，夏季移往山区，秋季再迁往平原。13 世纪的部分土库曼人、卡尔梅克-蒙古人、蒙古人就曾践行此种游牧模式。第二种是"山区—平原"型：冬季住山区，夏季移往河、湖边放牧。许多东部哈萨克牧民遵循这种方式。第三种是"山区—山脚—山区"型：冬季在山区避风处，春季移往山脚，夏季又往山区放牧，秋季下降至离春草场不远的地方，冬季再回到山区。萨彦岭地区的图瓦牧民、部分蒙古国与阿尔泰山牧民以及多数吉尔吉斯牧民，皆采用此种游牧模式。第四种为"山区"型：夏季在接近山脊处放牧，冬季下降到山谷森林中，整年不离山区，这是东部图瓦驯鹿牧人的游牧方式。

According to soviet ethnographer Sevyan Vainshtein, the "Turko-Mongol" herders of the Eurasian steppe can be divided into four types. The first one is the "plain-mountain-plain" type: they inhabit plains in the winter, moving to mountains in the summer and again to plains in the autumn. This pattern of nomadism was practiced by some of the Turkmen, Kalmyk-Mongols, and a part of the Mongols in the 13th century. The second one is the "mountain-plain" type: they inhabit mountains in the winter, moving to riverside or lakeside in the summer. Many nomads of the eastern Kazakh tribes followed this pattern. The third one is the "mountain-mountain foot-mountain" type: they spend the winter in the mountains sheltered from the wind, move to the foothills in the spring, and then head back to the mountains to graze in the summer. In the autumn, they will descend to meadows, returning to mountains in the winter. This pattern was followed by the Tuvan nomads of the Sayan mountains, some of the Mongol nomads and Altai nomads, as well as the majority of Kyrgyz nomads. The fourth one is the "mountain" type: they herd their cattle along the ridge of mountains, descending to forested valleys in the winter, which means that their nomadism never goes beyond mountains. This is the way by which the eastern Tuvan nomads herd their reindeers.

中国制造的男式长袍

布里亚特人，后贝加尔湖地区 ｜ 19 世纪中叶 ｜ 棉织物，丝绸，天鹅绒，白色合金，金线

Male robe (jacket) produced in China

Buryats. Transbaikal Oblast ｜ Mid-19th century ｜ Cotton, silk, velvet, white alloy, gold thread

女士无袖外套

布里亚特人，后贝加尔湖地区 ｜ 19 世纪末 ~ 20 世纪初 ｜ 丝绸，棉织物，天鹅绒，白色合金

Woman sleeveless jacket

Buryats. Transbaikal Oblast ｜ Late 19th – early 20th century ｜ Silk, cotton, velvet, white alloy

已婚女士长裙

喀尔喀蒙古人，蒙古国，库伦市 | 19 世纪末 | 丝绸，棉织物，铜合金，锦缎，金属线

Married woman dress

Khalkha-Mongols. Mongolia, Urga | Late 19th century | Silk, cotton, copper alloy, brocade, metal thread

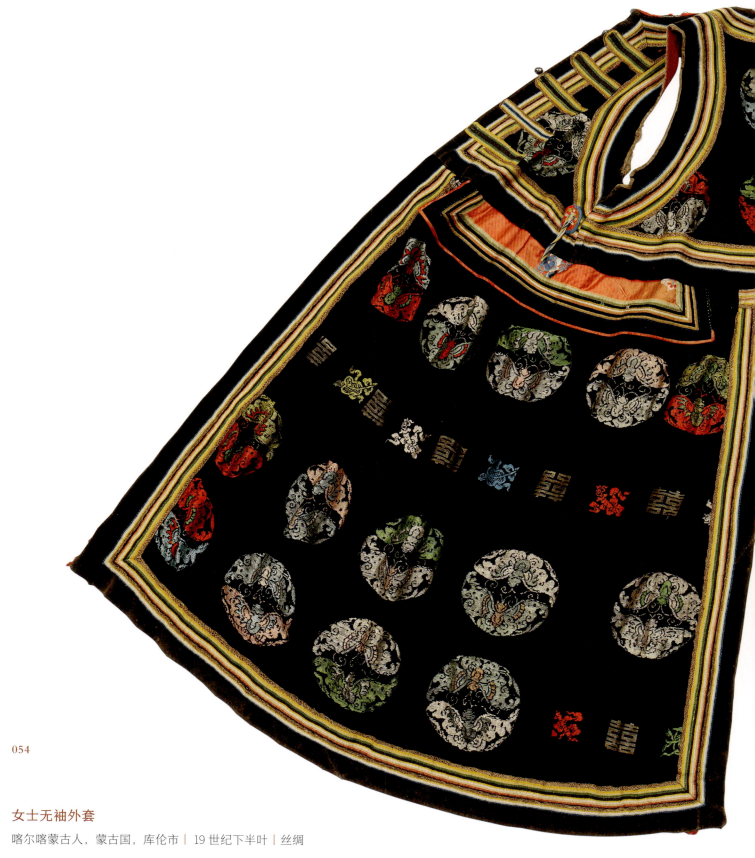

女士无袖外套

喀尔喀蒙古人，蒙古国，库伦市 ｜ 19 世纪下半叶 ｜ 丝绸

Woman sleeveless jacket

Khalkha-Mongols. Mongolia, Urga ｜ Second half of the 19th century ｜ Silk

056

一位富裕男士的节日服装
图瓦人
19 世纪末 ~ 20 世纪初
Festive costumes of a wealthy man
Tuvans
Late 19th ~ early 20th century

男式节日长袍

图瓦人，乌梁海边疆区 | 19 世纪末 | 丝绸，天鹅绒，棉织物，铜合金

Male festive robe

Tuvans. Uryankhay Kray | Late 19th century | Silk, velvet, cotton, copper alloy

058

一位富裕女士的节日服装
图瓦人
19 世纪末 ~ 20 世纪初
Festive costumes of a wealthy woman
Tuvans
Late 19th - early 20th century

女士节日长袍

图瓦人，乌梁海边疆区 | 19 世纪末 | 丝绸，天鹅绒，棉织物，铜合金，金属丝

Woman festive robe

Tuvans. Uryankhay Kray | Late 19th century | Silk, velvet, cotton, copper alloy, metal thread

年轻摔跤手服装

图瓦人，图瓦苏维埃社会主义自治共和国 | 20 世纪 70 年代 | 丝绸

Young wrestler attire

Tuvans. Tuvan ASSR | 1970s | Silk

女士无袖外套

阿尔泰人，托木斯克省，比斯克区 ｜ 19 世纪末 ~ 20 世纪初 ｜ 锦缎，丝线，金线，树脂

Woman sleeveless jacket

Altay-kiji. Tomsk Province, Biysk District ｜ Late 19th - early 20th century ｜ Brocade, silk thread, gold thread, mastic

女士无袖外套

哈卡斯人，叶尼塞省，米努辛斯克市 | 19 世纪晚期 | 丝绸，长毛绒，棉织物，丝线，金线

Woman sleeveless jacket

Khakas. Yenisei Province, Minusinsky Uyezd | Late 19th century | Silk, plush, cotton, silk thread, gold thread

装饰性小包（女士皮草配饰）

哈卡斯-克孜勒兹人，叶尼塞省，阿钦斯克地区

19 世纪末 ~ 20 世纪初

丝绸，长毛绒，棉织物，丝线

Decorative pouch – adornment for the woman furcoat

Khakas-kyzy, Yenisei Province, Achinsky Uyezd

Late 19th - early 20th century

Silk, plush, cotton, silk thread

手套

哈卡斯人，叶尼塞省，米努辛斯克市 ｜ 19 世纪末 ~ 20 世纪初 ｜ 布，丝绸，丝线，绸缎，棉线

Gloves

Khakas. Yenisei Province, Minusinsky Uyezd ｜ Late 19th - early 20th century ｜ Cloth, silk, silk thread, satin, cotton thread

烟袋

哈卡斯–科钦人，叶尼塞省，米努辛斯克市
19 世纪末 ~ 20 世纪初
丝绸，棉织物，长毛绒，丝线，玻璃

Tobacco pouch

Khakas–katchins. Yenisei Province, Minusinsky Uyezd
Late 19th – early 20th century
Silk, cotton, plush, silk thread, glass

064

烟袋

哈卡斯–科钦人，叶尼塞省，米努辛斯克市 | 19 世纪末 ~ 20 世纪初 | 长毛绒，棉织物，丝绸，丝线，棉线

Tobacco pouch

Khakas–katchins. Yenisei Province, Minusinsky Uyezd | Late 19th – early 20th century | Plush, cotton, silk, silk thread, cotton thread

女士节日长袍

哈卡斯人，哈卡斯自治区 | 20 世纪中叶 | 丝绸，长毛绒，布，绸缎，丝线，棉线

Woman festive kaftan (robe)

Khakas. Khakas Autonomous Oblast | Mid-20th century | Silk, plush, cloth, satin, silk thread, cotton thread

女士节日长裙

哈卡斯人，哈卡斯自治区，阿巴坎 | 1958 年 | 丝绸，羊毛，塑料，棉织物，丝线，棉线

Woman festive dress

Khakas. Khakas Autonomous Oblast, Abakan | 1958 | Silk, wool, plastic, cotton, silk thread, cotton thread

女士节日服装
卡尔梅克人
20 世纪初
Woman festive costumes
Kalmyks
Early 20th century

已婚妇女头饰

卡尔梅克人，阿斯特拉罕省

19 世纪末 ~ 20 世纪初

丝绸，长毛绒，锦缎，金银绦带，金线，丝线

Married woman headwear

Kalmyks. Astrakhan Province

Late 19th – early 20th century

Silk, plush, brocade, passementerie, gold thread, silk thread

068

女士长裙

卡尔梅克人，顿河哥萨克地区 │ 19 世纪中叶 │ 丝绸，棉织物，半锦缎

Woman dress

Kalmyks. Don Host Oblast │ Mid-19th century │ Silk, cotton, semi-brocade

女士节日无袖外套

卡尔梅克人，阿斯特拉罕省 | 19 世纪末 ~ 20 世纪初 | 波纹织物，金银绦带，玻璃，金线，丝线，铜合金

Woman festive sleeveless jacket

Kalmyks. Astrakhan Province | Late 19th – early 20th century | Silk (moire), passementerie, glass, gold thread, silk thread, copper alloy

女士节日无袖外套

卡尔梅克人，阿斯特拉罕省 | 19 世纪末 | 绸缎，天鹅绒，棉织物，羊毛线，白色合金

Woman festive sleeveless jacket

Kalmyks. Astrakhan Province | Late 19th century | Satin, velvet, cotton, wool thread, white alloy

女士披肩

卡尔梅克人，阿斯特拉罕省 ｜ 20 世纪初 ｜ 丝绸，丝线

Woman kerchief

Kalmyks. Astrakhan Province ｜ Early 20th century ｜ Silk, silk thread

女士长袍

布里亚特人，布里亚特苏维埃社会主义自治共和国，乌兰乌德市 ｜ 20 世纪 80 年代初 ｜ 丝绸，棉织物，天鹅绒，绸缎，白色合金

Woman robe

Buryats. Buryat ASSR, Ulan-Ude ｜ Early 1980s ｜ Silk, cotton, velvet, satin, white alloy

女士节日头饰

布里亚特人，赤塔州，阿金斯克-布里亚特自治区

21 世纪初

丝绸，金线，丝线，锦缎，天鹅绒，厚呢子

Woman festive headwear

Buryats. Chitinskaya Oblast, Agin-Buryat Autonomous Okrug

Early 21st century

Silk, gold thread, silk thread, brocade, velvet, drape cloth

女士无袖外套

布里亚特人，布里亚特苏维埃社会主义自治共和国，乌兰乌德市 | 20 世纪 80 年代初 | 丝绸，棉织物，白色合金

Woman sleeveless jacket

Buryats. Buryat ASSR, Ulan-Ude | Early 1980s | Silk, cotton, white alloy

受中国影响地区的佛教用品
Buddhist Objects in China-influenced Areas

在中国影响范围内的地区，丝线、线绳和织物等丝织品不仅用于日常文化生活，还被广泛用于制作佛教僧侣服饰、寺庙内部装饰和佛教礼仪用品。这些事实表明，丝绸具有高度完美的艺术性和尊贵的地位。各种颜色和质地的丝绸、锦缎织物以及丝线都可以作为独立的材料，用于制作贴花和刺绣佛像。象征着北传佛教文化的哈达，是一种使用主要佛教颜色（白、黄、蓝、绿、红）制作的丝绸织物，不仅用于佛教礼仪活动，还会作为礼物馈赠。

In the territories lying within China's sphere of influence silk pieces like thread, cords and fabrics were used not only in everyday life and culture. These textile items were widely applied for manufacturing costume elements for the Buddhist clergy, decorating temple interiors and designing various ceremonial and ritualistic objects. Such facts point to the highly developed aesthetic and prestigious status of the material. Silk thread, silk and brocade fabrics of various colors and textures could be used as independent materials in making applique and embroidering Buddhist images. The khadaks, which symbolises the culture of the Northern type of Buddhist peoples, is a silk fabric made using the main Buddhist colours (white, yellow, blue, green and red). This fabric is not only used in Buddhist ceremonial and ritualistic practices, but also given as gifts.

佛教圣像"曼陀罗"

布里亚特人，后贝加尔湖地区 | 19 世纪末 | 矿物颜料，丝绸，画布，木

Buddhist icon "Mandala"

Buryats. Transbaikal Oblast | Late 19th century | Mineral pigments, silk, canvas, wood

寺庙室内装饰（也在仪式游行时使用）

布里亚特人，后贝加尔湖地区

19 世纪末

丝绸，天鹅绒，棉织物，锦缎，金银绦带，丝线，

铜合金，白色合金，木

**Interior decoration of the Buddhist
temple (The items are also used as
attributes of the ritual processions)**

Buryats. Transbaikal Oblast

Late 19th century

Silk, velvet, cotton, brocade, passementerie,

silk thread, copper alloy, white alloy, wood

077

佛坛桌布

蒙古人，蒙古国 │ 20 世纪初 │ 丝绸，棉织物

Altar cloth

Mongols. Mongolia │ Early 20th century │ Silk, cotton

佛教僧侣服装
蒙古人
20 世纪初
Buddhist monk costumes
Mongols
Early 20th century

佛教僧侣节日长袍

喀尔喀蒙古人，蒙古国，库伦市 | 20 世纪初 | 丝绸，棉织物，铜合金

Buddhist monk festive robe

Khalkha-Mongols. Mongolia, Urga | Early 20th century | Silk, cotton, copper alloy

佛教僧侣头饰

喀尔喀蒙古人，蒙古国，库伦市
20 世纪初
丝绸，天鹅绒，棉织物

Buddhist monk headwear

Khalkha-Mongols. Mongolia, Urga
Early 20th century
Silk, velvet, cotton

羊毛腰带

布里亚特人 | 20 世纪初 | 羊毛

Wool belt

Buryats | Early 20th century | wool

佛教僧侣外衣

图瓦人，乌梁海边疆区 | 19 世纪末 | 丝绸，天鹅绒，棉织物，铜合金，棉线

Buddhist monk outerwear

Tuvans. Uryankhay Kray | Late 19th century | Silk, velvet, cotton, copper alloy, cotton thread

高级佛教僧侣（喇嘛）长袍

卡尔梅克人，斯塔夫罗波尔省｜19 世纪下半叶｜丝绸，棉织物

Buddhist monk (of a high rank) robe

Kalmyks. Stavropol Province｜Second half of the 19th century｜Silk, cotton

佛教僧侣外套

布里亚特人，后贝加尔湖地区，阿金斯基地区 | 19 世纪末 ~ 20 世纪初 | 丝绸，半毛织物，棉织物

Buddhist monk jacket

Buryats. Transbaikal Oblast, Aginsky District | Late 19th - early 20th century | Silk, semi-wool, cotton

俄罗斯远东地区的丝绸
Silk in the Russian Far East

　　与中国接壤的俄罗斯远东地区是另一个将丝绸和丝绸刺绣作为本民族文化中不可分割的一部分的地区。该地区包括那乃族、乌尔奇族、乌德盖族、乌尔塔族、涅吉达尔族、奥罗奇族、尼夫赫族和阿伊努族。这些民族的传统服饰都以多种元素组合为特点。除了东亚风格的长袍、帽饰和鞋子外，还有长裤、短裤、皮靴、长袜、袖套、手套、胸衣、可拆卸衣领、衬衫、马甲、耳包、围裙、腰带和其他配饰。

　　尤其值得一提的是那乃族新娘的礼服，衣服的前胸和后背处可以制作成不同的款式，比如，前胸采用贴花技术，用对比鲜明的布片做成龙鳞的形状，后背则使用底色材料和缎面刺绣绣上对称的传统装饰图案"生命之树"，树下是鱼和两栖动物的形象，树上则是隐藏在树叶中的鸟。

　　除了传统服装，远东地区的人们还用丝绸制作床上用品。19世纪开始，人们使用丝绸制作图案绚丽的壁毯。在现代文化中，经过女织工精巧的手艺加工，壁毯逐渐成为值得博物馆收藏的艺术品。

One more border territory with China is the Far East. It is the region where silk fabric and embroidery with silk thread became an integral part of the culture of indigenous peoples. The region includes the Nanai, the Ulch, the Udeghe, the Uilta, the Negildas, the Orochi, the Nivkhs and the Ainu. Traditional costumes of these peoples are characterized by pronounced multielement. Besides Eastern Asia types of robes, headwear and shoes, they used trousers or trunks, high gaiters, stockings, sleevelets, gloves, breastplates, detachable collars, shirts, sleeveless jackets, headgear, aprons, belts and other accessories.

One more interesting piece is a Nanai bride's robe. The front and back parts could be made in different techniques and styles. For example, the front parts were made of various contrast patches in the appilque technique which mimics the form of a dragon scale. On the back, a traditional symmetrical ornament, "tree of life", is embroidered using sole-colored material and satin embroidery, with images of fishes and amphibians underneath the tree, and birds hidden in the leaves above the tree.

Except for traditional costumes, people of the Far East region used silk for making bedding. Since the 19th century they have decorated wall-hanging carpets with complicated ornamental compositions. In modern culture, women weavers turn such hangings into pieces of art that worth being in a museum's collection.

那乃族
Nanai

　　那乃族（也称赫哲族）是东亚的通古斯民族，居住在黑龙江、松花江和乌苏里江中游的阿穆尔盆地一带。那乃族的祖先是女真人。

The Nanai people are a Tungusic people of East Asia who have traditionally lived along Heilongjiang (Amur), Songhuajiang (Sunggari) and Wusuli River (Ussuri) in the Middle Amur Basin. The ancestors of the Nanai were the Jurchens of northernmost Manchuria (outside China- Russian Manchuria).

那乃族萨满巫师（拍摄于19世纪末）
Nanai shamans, photographed at the end of the 19th century

新娘长袍

那乃人，远东边疆地区 | 20 世纪初 | 丝绸，棉织物，丝线，棉线，金线

Bride's robe

Nanai. Far Eastern Krai | Early 20th century | Silk, cotton, silk thread, cotton thread, gold thread

新娘头饰

那乃人，远东边疆地区
20 世纪初
丝绸，棉织物，棉线

Bride's hearwear

Nanai. Far Eastern Krai
Early 20th century
Silk, cotton, cotton thread

猎人斗篷兜帽

那乃人，远东边疆地区 | 19 世纪末 ~ 20 世纪初 | 丝绸，棉织物，丝线，棉线，铜合金

Hunter's head cover

Nanai. Far Eastern Krai | Late 19th – early 20th century | Silk, cotton, silk thread, cotton thread, copper alloy

男式长袍

那乃人，远东边疆地区｜19 世纪末～20 世纪初｜布，棉织物，羊毛织布，丝线，棉线，铜合金，玻璃

Male robe

Nanai. Far Eastern Krai ｜ Late 19th – early 20th century ｜ Cloth, cotton, woolen cloth, silk thread, cotton thread, copper alloy, glass

护胫

那乃人，远东边疆地区
19 世纪末
丝绸，棉织物，丝线，棉线

Greaves

Nanai. Far Eastern Krai
Late 19th century
Silk, cotton, silk thread, cotton thread

088

男式腰带

那乃人，远东边疆地区 ｜ 19 世纪末 ~ 20 世纪初 ｜ 棉织物，棉线，丝线

Male belt

Nanai. Far Eastern Krai ｜ Late 19th – early 20th century ｜ Cotton, cotton thread, silk thread

男式腰包

那乃人，远东边疆地区
19 世纪末 ~ 20 世纪初
棉织物，丝线

Male belt pouch

Nanai. Far Eastern Krai
Late 19th – early 20th century
Cotton, silk thread

089

枕套

那乃人，哈巴罗夫斯克边疆地区｜20 世纪上半叶｜棉织物，羊毛，丝线，棉线

Pillow case

Nanai. Khabarovsk Krai ｜ First half of the 20th century ｜ Cotton, wool, silk thread, cotton thread

男式节日长袍

乌德盖人，哈巴罗夫斯克边疆地区 | 20 世纪中叶 | 丝绸，棉织物，丝线，棉线，铜合金

Male festive robe

Udege. Khabarovsk Krai | Mid-20th century | Silk, cotton, silk thread, cotton thread, copper alloy

新娘长袍

奥罗奇人，远东边疆地区 | 19 世纪末 ~ 20 世纪初 | 丝绸，棉织物，丝线，金线，棉线，铜合金

Bride's robe

Orochi. Far Eastern Krai | Late 19th – early 20th century | Silk, cotton, silk thread. gold thread, cotton thread, copper alloy

男式臂章

奥罗奇人，远东边疆地区

19 世纪末

棉织物，绸缎，丝线，棉线

Male cuffs

Orochi. Far Eastern Krai

Late 19th century

Cotton, satin, silk thread, cotton thread

092

儿童围兜

奥罗奇人，远东边疆地区 ｜ 19 世纪末 ~ 20 世纪初 ｜ 丝绸，棉织物，棉线

Children breast decoration

Orochi. Far Eastern Krai ｜ Late 19th – early 20th century ｜ Silk, cotton, cotton thread

女士葬礼护胸装饰

奥罗奇人，远东边疆地区

19 世纪末

丝绸，棉织物，丝线，棉线

Woman funeral breast decoration

Orochi. Far Eastern Krai

Late 19th century

Silk, cotton, silk thread, cotton thread

093

男式葬礼围裙

奥罗奇人，远东边疆地区 ┃ 19 世纪末 ┃ 丝绸，棉织物，棉线

Male funeral apron

Orochi. Far Eastern Krai ┃ Late 19th century ┃ Silk, cotton, cotton thread

3

中亚文化中
的丝绸

Silk
in the Culture of
Central Asia

染色前在经线上绘制图案
乌兹别克人，撒马尔罕
1902年
Patterning on the wrap yarn before dyeing
Uzbeks. Samarkand
1902

丝绸加工车间
乌兹别克人，撒马尔罕
1902年
Silk reeling workshop
Uzbeks. Samarkand
1902

费尔干纳绿洲的养蚕业
Sericulture of the Fergana Oasis

　　自古以来，费尔干纳盆地是欧亚大陆最大的蚕桑中心之一。该地具备丝绸织造得天独厚的自然条件，而且地处丝绸之路的要道上，这也唤醒了人们对丝绸纺织业的兴趣。20 世纪初，费尔干纳绿洲的丝织品包括各种类型的丝绸和半丝绸品，其中包括缎织品。中亚的其他丝织中心则不生产缎织品。

　　费尔干纳丝绸的传统还包括生产轻薄的"杜鲁亚"面料，这种面料每根线只有 4 根丝纤维。这种面料的生丝不需要像其他类型的丝绸那样进行煮沸，但织物本身是经过煮沸的，并采用木版印刷或扎染加以装饰。这项丝织技术由梅尔夫市的移民织工传播到费尔干纳盆地。"杜鲁亚"面料与东北高加索地区的"克拉加依"面料相似。

From ancient times, the Fergana Valley was one of the largest sericulture centers in Eurasia. It had all the necessary natural conditions for silk farming, and the location on the main route of the Silk Road contributed to the growing popularity of this occupation. In the early 20th century, the Fergana Oasis production of silk fabrics offered a variety of silk and semi-silk fabrics, including satin ones. Satins were exclusive to the region and were not produced in other silk centers of Central Asia.

The Fergana Oasis specialized in manufacturing a thin lightweight fabric of duriya which had only 4 silk fibers per thread. The raw silk material for this fabric was not boiled, unlike other types of silk, but the fabric itself was, and decorated with woodblock-printed or tie-dyed. This technology was brought to the Fergana Valley by the weavers from Merv, duriya is similar to the fabric of Kragai common across the North-Eastern Caucasus.

费尔干纳盆地
The Fergana Valley

中亚的费尔干纳盆地位于乌兹别克斯坦东部、吉尔吉斯斯坦南部和塔吉克斯坦北部。费尔干纳盆地是位于中亚的一个山间洼地，处于北部的天山山系和南部的阿莱山系之间。谷地长约300公里，宽达70公里，面积达22000平方公里。它的位置使其成为一个独立的地理区域。纳伦河和卡拉河这两条河流在纳曼干附近的河谷中汇合，形成锡尔河。

费尔干纳盆地的历史可以追溯到2300多年前，亚历山大大帝在其西南部建立了亚历山大城。这里曾是希腊、中国、巴克特里亚和帕提亚文明之间的通道。莫卧儿王朝的缔造者巴布尔曾在此居住，将该地区与现代阿富汗和南亚联系在一起。19世纪末，费尔干纳盆地成为俄罗斯帝国的一部分，20世纪20年代成为苏联的一部分。该地区大部分是穆斯林居住区，这里居住着乌孜别克族、塔吉克族和吉尔吉斯族，他们通常混居在一起。

The Fergana Valley in Central Asia lies mainly in eastern Uzbekistan, but also extends into southern Kyrgyzstan and northern Tajikistan. The Fergana Valley is an intermountain depression in Central Asia, between the mountain systems of the Tien Shan in the north and the Alay in the south. The valley is approximately 300 kilometres long and up to 70 kilometres wide, forming an area covering 22,000 square kilometres. Its position makes it a separate geographic zone. The valley owes its fertility to two rivers, the Naryn and the Kara Darya, which unite in the valley, near Namangan, to form the Syr Darya.

The valley's history stretches back over 2,300 years, when Alexander the Great founded Alexandria Eschate at its southwestern end, as a path between Greek, Chinese, Bactrian and Parthian civilisations. It was home to Babur, the founder of the Mughal dynasty, tying the region to modern Afghanistan and South Asia. The Russian Empire conquered the valley at the end of the 19th century, and it became part of the Soviet Union in the 1920s. The area largely remains Muslim, populated by ethnic Uzbek, Tajik and Kyrgyz people, often intermixed.

女士长袍

萨尔特人，费尔干纳省 | 19 世纪 70 ~ 90 年代 | 丝绸，棉线，金属（捻线），丝线

Woman robe

Sart. Fergana Oblast | 1870-1890s | Silk, cotton thread, metal (twisted thread), silk thread

丝绸织物

萨尔特人，费尔干纳省，科坎德市｜19 世纪 70 ~ 90 年代｜丝绸

Silk fabric

Sart. Fergana Oblast, Kokand｜1870-1890s｜Silk

丝绸织物

萨尔特人，费尔干纳省，科坎德市

1902 年

丝绸，捻丝

Silk fabric

Sart. Fergana Oblast, Kokand

1902

Silk, twisted silk

丝绸织物

萨尔特人，费尔干纳省，科坎德市

1902 年

丝绸，捻丝

Silk fabric

Sart. Fergana Oblast, Kokand

1902

Silk, twisted silk

丝绸织物

萨尔特人，费尔干纳省，科坎德市 ｜ 1902 年 ｜ 丝绸，捻丝

Silk fabric

Sart. Fergana Oblast, Kokand ｜ 1902 ｜ Silk, twisted silk

丝绸织物

萨尔特人，费尔干纳省，科坎德市

19 世纪 70 ~ 90 年代

丝绸

Silk fabric

Sart. Fergana Oblast, Kokand

1870-1890s

Silk

丝绸织物

萨尔特人，费尔干纳省，科坎德市

19 世纪 60 ~ 90 年代

丝绸，捻丝

Silk fabric

Sart. Fergana Oblast, Kokand

1860-1890s

Silk, twisted silk

103

女士套装
乌兹别克人
19 世纪 70 ~ 90 年代
Woman attire
Uzbeks
1870-1890s

衬衫

乌兹别克人，费尔干纳省，科坎德市 | 19 世纪 70 ~ 90 年代 | 半丝织物，丝绸，丝线

Shirt

Uzbeks. Fergana Oblast, Kokand | 1870–1890s | Semi-silk, silk, silk thread

女士长裤

乌兹别克人，突厥斯坦 | 19 世纪 60 年代 | 丝绸，印花布

Woman trousers

Uzbeks. Turkestan | 1860s | Silk, calico

头巾

乌兹别克人，费尔干纳省，科坎德市
19 世纪 70 ~ 90 年代
丝绸

Head kerchief

Uzbeks. Fergana Oblast, Kokand
1870–1890s
Silk

106

女士头巾

乌兹别克人，布哈拉市 | 20 世纪初 | 丝绸

Woman head kerchief

Uzbeks. Bukhara | Early 20th century | Silk

项链

乌兹别克人，突厥斯坦

19 世纪 60 年代

胶剂，红珊瑚，黄色金属（箔），玻璃

Necklace

Uzbeks. Turkestan

1860s

Paste, red coral, yellow foil, glass

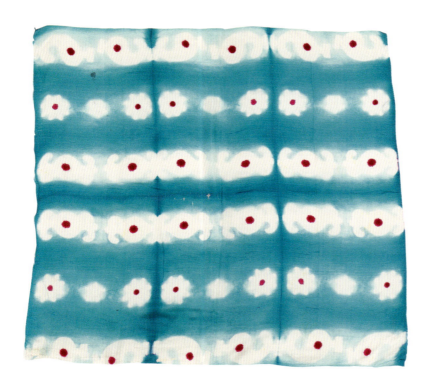

披巾

乌兹别克人，乌兹别克苏维埃社会主义共和国，塔什干 | 1954 年 | 丝绸

Scarf

Uzbeks. Uzbek SSR, Tashkent | 1954 | Silk

女士头巾

乌兹别克人，希瓦汗国 | 1913 ~ 1915 年 | 丝绸

Woman head kerchief

Uzbeks. Khanate of Khiva | 1913–1915 | Silk

丝绸织物

乌兹别克人，中亚 | 1875 ~ 1900 年 | 丝绸

Silk fabric

Uzbeks. Central Asia | 1875-1900 | Silk

男式长袍

乌兹别克人，布哈拉酋长国，布哈拉市 ｜ 1875 ～ 1900 年 ｜ 丝绸，丝线，铜

Male robe

Uzbeks. Emirate of Bukhara, Bukhara ｜ 1875-1900 ｜ Silk, silk thread, copper

男式长裤

乌兹别克人，中亚和哈萨克斯坦，突厥斯坦｜1860～1870 年｜棉织物，丝绸，丝线，棉线

Male trousers

Uzbeks. Central Asia and Kazakhstan, Turkestan｜1860-1870｜Cotton, silk, silk thread, cotton thread

面纱

塔吉克高山人，达尔瓦扎山 | 19 世纪 60 ~ 80 年代 | 棉织物，丝线

Face veil

Mountain (or Chinese) Tajiks. Mt Darwaza | 1860–1880s | Cotton, silk thread

面纱

塔吉克高山人，达尔瓦扎山 | 19 世纪 60 ～ 80 年代 | 棉织物，丝线

Face veil

Mountain (or Chinese) Tajiks. Mt Darwaza | 1860-1880s | Cotton, silk thread

面纱

塔吉克高山人，达尔瓦扎山 | 19 世纪末 | 棉织物，丝线

Face veil

Mountain (or Chinese) Tajiks. Mt Darwaza | Late 19th century | Cotton, silk thread

布哈拉和撒马尔罕地区的丝绸织物
Silk Fabrics in Bukhara and Samarkand

　　早在公元一千纪中期，泽拉夫尚河中游的绿洲（现为乌兹别克斯坦共和国领土）以丝织品闻名于世，当时该地区被称为索格狄亚纳。在绿洲最大的纺织中心——布哈拉和撒马尔罕，数百年来一直保留着精湛的丝绸纺织传统，甚至在 20 世纪初，这里仍在生产高质量和具有杰出艺术价值的手工丝织品。这些织物主要是"依卡特"面料，中亚人称之为"阿布尔"，这些织物采用经纱防染的方法进行装饰。

　　这些织物主要用于缝制中亚社会精英的服装以及节日装饰等。中亚丝绸在布哈拉酋长国的游牧民族中很受欢迎，被视为名贵的消费品。而在中亚以外，"依卡特"布料在伏尔加河，以及乌拉尔和西伯利亚地区的鞑靼人当中非常畅销。"依卡特"布料的风格化图案经常被用来装饰俄罗斯工厂的纺织品。

The oasis in the middle of the Zeravshan River (now the territory of the Republic of Uzbekistan) was famous for its silk fabrics back in the mid-first millennium, and at that time the region was known as Sogdiana. Traditions of elite silk weaving were preserved over centuries in the largest textile centers of the oasis – Bukhara and Samarkand, where in the early 20th century the manufacture of artisanal silk fabrics of high quality and outstanding artistic merits never stopped. Bukhara and Samarkand were famous for the ikat fabrics, which in Central Asia were known as abr. They were decorated with a method of reserve dyeing of the base.

These fabrics were mainly meant for the elite's garments of the Central Asian society, festive caparisons, etc. The Central Asian silk was popular with the nomadic population of the Emirate of Bukhara as prestigious goods. Outside Central Asia, ikats were popular with the Tatars of the Volga Region and the Ural Region, as well as with the Siberian Tatars. Stylized images of ikat fabrics were often used in the ornaments of Russian textile manufacturers.

索格狄亚纳和索格狄亚纳人
Sogdiana and Sogdians

　　索格狄亚纳是一支位于阿姆河和锡尔河之间的古代伊朗文明，位于今天的乌兹别克斯坦、土库曼斯坦、塔吉克斯坦、哈萨克斯坦和吉尔吉斯斯坦境内。索格狄亚纳曾是阿契美尼德帝国的一个行省，被记录在大流士一世的贝希斯敦铭文之上。索格狄亚纳首先被阿契美尼德帝国的缔造者居鲁士大帝征服，后于公元前 328 年被马其顿统治者亚历山大大帝吞并。在塞琉古帝国、希腊—巴克特里亚王国、贵霜帝国、萨珊帝国、嚈哒帝国、西突厥汗国以及穆斯林征服河中地区期间，索格狄亚纳的统治权不断易手。

　　索格狄亚纳城邦虽然在政治上从未统一过，但都以撒马尔罕城为中心。索格狄亚纳领土相当于现代乌兹别克斯坦的撒马尔罕和布哈拉地区，以及现代塔吉克斯坦的苏格德地区。

Sogdiana was an ancient Iranian civilization between the Amu Darya and the Syr Darya, and in present-day Uzbekistan, Turkmenistan, Tajikistan, Kazakhstan, and Kyrgyzstan. Sogdiana was also a province of the Achaemenid Empire, and listed on the Behistun Inscription of Darius the Great. Sogdiana was first conquered by Cyrus the Great, the founder of the Achaemenid Empire, and then was annexed by the Macedonian ruler Alexander the Great in 328 BCE. It would continue to change hands under the Seleucid Empire, the Greco-Bactrian Kingdom, the Kushan Empire, the Sasanian Empire, the Hephthalite Empire, the Western Turkic Khaganate and the Muslim conquest of Transoxiana.
The Sogdian city-states, although never politically united, were centered on the city of Samarkand. Sogdian territory corresponds to the modern regions of Samarkand and Bukhara in modern Uzbekistan, as well as the Sughd region of modern Tajikistan.

索格狄亚纳人（粟特人）是伊朗裔人，其发源地索格狄亚纳位于数条重要商路的中心，地处今天的乌兹别克斯坦和塔吉克斯坦。索格狄亚纳作为阿契美尼德帝国的一个行省，最早见于公元前5世纪的记载，后来被亚历山大大帝征服。在此期间，索格狄亚纳由一片片绿洲城镇和富饶的农田组成，处于亚洲大陆各帝国之间，地理位置得天独厚。

索格狄亚纳人在商旅途中建立起商人社区，在不同帝国势力的庇护下生活。他们拥有军队，参与军事行动，建立城防，以抵御来自草原的游牧民族对他们财富的不断威胁。

索格狄亚纳人的影响力中最令人惊讶的一点是，它并不依赖于政治或军事力量。他们没有建立帝国，其政治组织是一系列小公国，每个公国都有自己的首领。

索格狄亚纳人不仅熟悉各种外来文化和语言，还擅长经营丝绸和其他珍贵商品，包括费尔干纳盆地的马匹、印度的宝石、西藏的麝香、北方草原的毛皮等。索格狄亚纳人也是能工巧匠，他们制作并销售奢华的物品，尤其是金属制品和纺织品，足迹遍布亚洲草原和中国。

唐代三彩釉索格狄亚纳人陶俑
中国国家博物馆藏
Tang dynasty tri-colour glazed sogdiana terracotta figure (Collection of the National Museum of China)

唐代三彩釉陶载乐骆驼 中国国家博物馆藏
Tang dynasty tri-colour glazed earthenware figure of musicians on a camel (Collection of the National Museum of China)

The Sogdians were an Iranian people whose homeland, Sogdiana, was located at the center of several connecting trading routes, in present-day Uzbekistan and Tajikistan. First recorded in the 5th century BCE as a province of the Achaemenid Persian Empire, and later conquered by Alexander the Great on his journey east across Asia. During this time, Sogdiana was made up of a patchwork of oasis towns and rich agricultural land, uniquely placed between the great empires of the Asian continent. Branching out from these oasis towns, the Sogdians set up merchant communities as they traveled, living under the aegis of various imperial powers. The people of Sogdiana certainly had armies, engaged in military campaigns, and fortified their towns to defend against the constant threat of nomads from the steppes who coveted the Sogdians' wealth.

One of the most surprising elements of the Sogdians' influence is that it did not rely on political or military power. They had no empire, and their political organization at home in Sogdiana was a series of small principalities, each with its own leader.

The Sogdians are not only familiar with a variety of foreign cultures and languages, but also specialise in the sale of silks and other precious commodities. Among these were horses from the Ferghana Valley, gemstones from India, musk from Tibet, and furs from the steppes to the north. The Sogdians were also skilled artisans, making and selling luxurious objects—particularly metalwork and textiles—across the Asian steppe and into China.

撒马尔罕
Samarkand

2500 多年来，撒马尔罕一直是世界文化的交汇点，也是穿越中亚的丝绸之路上最重要的节点之一。撒马尔罕位于乌兹别克斯坦东北部的泽拉夫尚河谷，拥有丰富的自然资源，该地区的定居历史可追溯到公元前 1500 年。

撒马尔罕长期以来一直是这个地区的贸易中心，在公元前 329 年被亚历山大征服之前的几个世纪，它是一座以手工艺品生产而闻名的大城市，拥有城堡和坚固的防御工事。从古代晚期到中世纪早期，索格狄亚纳人一直居住在这座城市及其周边地区。早在汉代，当中国人首次将他们对内亚地区的印象写成文字时，索格狄亚纳商人就被记录在中国人对该地区的描述中。公元 313 ~ 314 年发现的索格狄亚纳人书信，提供了有关撒马尔罕商人沿着贸易路线建立的商贸网络的证据。这些商人远行至中国各地，从事贵金属、香料和布匹贸易。巴基斯坦北部岩石上的索格狄亚纳铭文证明了他们南下印度时的活动。6 世纪，索格狄亚纳商人向西旅行，开辟了与拜占庭贸易的新路线。

8 世纪初穆斯林将该地区征服后，泽拉夫尚河谷许多小国的最后一批索格狄亚纳统治者纷纷逃亡。到 9 世纪，在伊朗裔萨曼王朝的统治下，河中地区的城市成为穆斯林的主要学习中心。1220 年，成吉思汗和他的蒙古军队征服了撒马尔罕，并将其摧毁，撒马尔罕的许多历史建筑都化为废墟。14 世纪 70 年代，撒马尔罕成为帖木儿王朝的首都。1500 年，撒马尔罕被并入布哈拉汗国。然而，随着布哈拉成为首都，撒马尔罕逐渐衰落，至 18 世纪末已无人居住，直到 1888 年铁路的开通才使这座城市重新焕发生机，恢复了昔日作为贸易中心的地位。时至今日，丝绸编织和纺织品的生产与贸易仍然是这座城市的主要产业之一。

The city of Samarkand has been at the crossroads of world cultures for over two and a half millennia, and is one of the most important sites on the Silk Routes traversing Central Asia. Located in the Zerafshan River valley, in north-eastern Uzbekistan, the city enjoys the benefits of abundant natural resources and settlement in the region can be traced back to 1500 BCE.

Samarkand has long been a central point for trade across the region, and was a substantial city renowned for its craft production, with a citadel and strong fortifications, several centuries before it was conquered by Alexander in 329 BCE. From the late antique and early medieval period, the city and the surrounding area were inhabited by the Sogdians. As early as Han times, when the Chinese first committed to writing their impressions of Inner Asia, Sogdian merchants were recorded in the Chinese descriptions of the region. Sogdian colonies were established all along the trade routes and Sogdian letters have been discovered from 313-314, providing evidence about a network of merchants from Samarkand, reaching various places as far as China, in order to trade precious metals, spices and cloth. Sogdian inscriptions on rocks in northern Pakistan testify to their activity on the routes south into India. Later on, in the 6th century, Sogdian merchants seemed to have travelled west and developed new routes for trade with Byzantium.

After the Muslim conquest in the early eighth century, the last of the Sogdian rulers of the many small states in the Zerafshan Valley fled. By the 9th century, the cities of Transoxiana became major centres of Muslim learning under the Samanids, who were of Iranian origin. Samarkand was invaded and destroyed by Genghis Khan and his Mongol armies when they conquered the area in 1220, and much of its historic architecture was reduced to ruins. In the 1370s, Samarkand became the capital of the Timurid dynasty. In 1500, Samarkand was incorporated into the Bukhara Khanate. Nonetheless, with Bukhara as the capital, Samarkand declined and was uninhabited by the late 18th century, only reviving with the introduction of the railway in 1888, which allowed the city to regain its ancient role as a trading centre at the crossroads of routes to east and west. Even today, silk weaving and the production and trade of textiles remain one of the city's major industries.

天鹅绒面料

萨尔特人，布哈拉市

19 世纪末

捻丝，棉织物

Velvet fabric

Sart. Bukhara

Late 19th century

Twisted silk, cotton

天鹅绒面料

萨尔特人，布哈拉市

19 世纪末

丝线，棉线

Velvet fabric

Sart. Bukhara

Late 19th century

Silk thread, cotton thread

丝绸面料

萨尔特人，布哈拉市

19 世纪末 ~ 20 世纪初

捻丝

Silk fabric

Sart. Bukhara

Late 19th – early 20th century

Twisted silk

天鹅绒面料

萨尔特人，布哈拉市

19 世纪末

捻丝，棉织物

Velvet fabric

Sart. Bukhara

Late 19th century

Twisted silk, cotton

丝绸面料

萨尔特人，布哈拉市

19 世纪末 ~ 20 世纪初

捻丝

Silk fabric

Sart. Bukhara

Late 19th – early 20th century

Twisted silk

丝绸面料

萨尔特人，布哈拉市

19 世纪末 ~ 20 世纪初

丝绸

Silk fabric

Sart. Bukhara

Late 19th – early 20th century

Silk

男式长袍

萨尔特人，撒马尔罕省｜19 世纪 70 ~ 90 年代｜棉织物，丝绸

Male robe

Sart. Samarkand Oblast ｜ 1870–1890s ｜ Cotton, silk

男式长袍

萨尔特人，布哈拉酋长国 | 19 世纪末 | 棉织物，半丝织物，绦带，印花布

Male robe

Sart. Emirate of Bukhara | Late 19th century | Cotton, semi-silk, narrow braid, calico

女士套装
萨尔特人
19 世纪末
Woman attire
Sart
Late 19th century

女士长裙领子装饰带
乌兹别克人，塔吉克人
19 世纪末
丝绸，金线，丝线
Woman Dress Collar Decoration Strap
Uzbeks. Tajik
Late 19th century
Silk, gold thread, silk thread

珠串项链

乌兹别克人，乌兹别克苏维埃社会主义共和国，
锡尔达利亚省，普沙格尔地区
20 世纪 80 年代
红珊瑚，棉线

Bead necklace

Uzbeks. Uzbek SSR, Sirdaryo Region, Pshagar
1980s
Red coral, cotton thread

头巾

萨尔特人，布哈拉酋长国，布哈拉市
19 世纪下半叶
丝绸，丝线，金属线

Head kerchief

Sart. Emirate of Bukhara, Bukhara
Second half of the 19th century
Silk, silk thread, metal thread

头巾

乌兹别克人，费尔干纳省，科坎德市
19 世纪 70 ~ 90 年代
丝绸

Head kerchief

Uzbeks. Fergana Oblast, Kokand
1870-1890s
Silk

128

女士外套

萨尔特人，布哈拉酋长国 ｜ 19 世纪末 ｜ 半丝织物，棉织物，丝绸

Woman jacket

Sart. Emirate of Bukhara ｜ Late 19th century ｜ Semi-silk, cotton, silk

已婚女士衬衫

萨尔特人，布哈拉酋长国

19 世纪末

半丝织物，棉织物，丝绸

Married woman shirt

Sart. Emirate of Bukhara

Late 19th century

Semi-silk, cotton, silk

女士长裤

萨尔特人，撒马尔罕省 ｜ 19 世纪末 ｜ 半丝织物，棉织物，棉线

Woman trousers

Sart. Samarkand Oblast ｜ Late 19th century ｜ Semi-silk, cotton, cotton thread

女士头巾

萨尔特人，布哈拉古城 | 19 世纪 70 ~ 90 年代 | 丝绸

Woman head kerchief

Sart. Old Bukhara | 1870-1890s | Silk

半丝织物 "艾德莱丝绸"

萨尔特人，撒马尔罕省，撒马尔罕市

1900 ~ 1902 年

丝绸，棉织物

Semi-silk fabric "adras"

Sart. Samarkand Oblast, Samarkand

1900–1902

Silk, cotton

半丝织物 "艾德莱丝绸"

萨尔特人，撒马尔罕省，撒马尔罕市

1900 ~ 1902 年

丝绸，棉织物

Semi-silk fabric "adras"

Sart. Samarkand Oblast, Samarkand

1900–1902

Silk, cotton

半丝织物 "艾德莱丝绸"

萨尔特人，撒马尔罕省，撒马尔罕市

1900 ~ 1902 年

丝绸，棉织物

Semi-silk fabric "adras"

Sart. Samarkand Oblast, Samarkand

1900–1902

Silk, cotton

花剌子模绿洲文化中的丝绸
Silk in the Culture of the Khwarezm Oasis

　　在中亚最古老的文明中心——花剌子模绿洲，曾出现过使用进口原料的丝绸纺织，但没有得到进一步发展。然而，该地区居民的节日服装中一定会采用进口丝绸。人们尤为青睐一种叫作"马达利"的织物，它由长条纹的红色丝绸制成。与此同时，该地区每个民族都形成了使用这种织物的独特传统：北花剌子模地区的乌兹别克人将其用作头巾，男性则用来做腰带；卡拉卡尔帕克的年轻人将它作为头巾披在头上。据资料记载，"马达利"是由土库曼部落制作的，历史上该部落与花剌子模联系密切。

In the Khwarezm Oasis, the oldest civilization center in Central Asia, silk weaving on imported raw materials though existed has not been much developed. However, the population of the oasis would always include items of the purchased silk as part of the festive attire. Madali was especially valued, it presented long streaks of red silk, weaved as piece goods. At the same time, each people of the oasis had their own silk tradition: the Uzbeks of Northern Khwarezm used it for their turbans, men everywhere used it for their belts, and the young Karakalpaks draped it over their heads as a wrapper. According to some sources, madali were made by Turkmen tribes, which are historically associated with Khwarezm.

花剌子模
Khwarezm

　　花剌子模位于中亚西部阿姆河三角洲的绿洲区域，是今天的土库曼斯坦、乌兹别克斯坦和哈萨克斯坦的一部分。它处在商队路线的十字路口，参与了丝绸之路上的文化交流。

　　公元 3 世纪至 8 世纪，花剌子模地区经历了多次政治变革。贵霜帝国衰落后，花剌子模地区政治上先后被不同的统治者控制，直到 7 世纪末阿拉伯人的到来。公元 4 世纪至 8 世纪，花剌子模与咸海地区、里海西北地区、伏尔加河地区和乌拉尔地区维持着密切的贸易关系。

Khwarezm is a large oasis region on the Amu Darya River delta in western Central Asia. Situated at the crossroads of the caravan routes, the Khwarezm region in modern Turkmenistan, Uzbekistan and Kazakhstan participated in the cultural exchanges and interactions that existed in the Silk Roads.

Between the 3rd and 8th centuries, the Khwarezm region experienced many political changes due to the circumstances of these times. After the decline of the Kushan Empire, the Khwarezm region witnessed issues between diverse rulings, until the arrival of the Arabs at the end of the 7th century. From the 4th to 8th century trading relations with the regions of the Aral Sea, the Northwest Caspian Sea area, the Volga, and Ural regions were significant for the economy of Khwarezm.

女士套装
卡拉卡尔帕克人
20 世纪初
Woman attire
Karakalpaks
Early 20th century

丝绸织物

卡拉卡尔帕克人，德尔塔阿姆河地区 ｜ 20 世纪初 ｜ 丝绸，玻璃

Silk fabric

Karakalpaks. Delta Amu Darya ｜ Early 20th century ｜ Silk, glass

女士头饰

卡拉卡尔帕克人，德尔塔阿姆河地区 | 19 世纪末 ~ 20 世纪初 | 羊毛，棉织物，羊毛织布，丝线

Woman headwear

Karakalpaks. Delta Amu Darya | Late 19th – early 20th century | Wool, cotton, woolen cloth, silk thread

女士长裙

卡拉卡尔帕克人，德尔塔阿姆河地区 ┃ 20 世纪初 ┃ 半丝织物，丝绸，丝线

Woman dress

Karakalpaks. Delta Amu Darya ┃ Early 20th century ┃ Semi-silk, silk, silk thread

女士长袍

卡拉卡尔帕克人，德尔塔阿姆河地区 | 19 世纪末 ~ 20 世纪初 | 半丝织物，棉织物，印花棉，绦带，植物纤维

Woman robe

Karakalpaks. Delta Amu Darya | Late 19th – early 20th century | Semi-silk, cotton, printed cotton, narrow braid, plant fiber

女士胸前装饰

卡拉卡尔帕克人，德尔塔阿姆河地区
19 世纪末
白色金属，红玛瑙石

Woman breast decoration

Karakalpaks. Delta Amu Darya
Late 19th century
White metal, cornelian

138

胸前装饰物

卡拉卡尔帕克人，德尔塔阿姆河地区 ｜ 19 世纪末 ~ 20 世纪初 ｜ 白色金属

Breast decoration

Karakalpaks. Delta Amu Darya ｜ Late 19th – early 20th century ｜ White metal

无边便帽

卡拉卡尔帕克人，卡拉卡尔帕克苏维埃
社会主义自治共和国
20 世纪 20 年代
长毛绒，丝线，印花棉，粘胶

Skull cap

Karakalpaks. Karakalpak ASSR
1920s
Plush, silk thread, printed cotton, glue

无边便帽

卡拉卡尔帕克人，卡拉卡尔帕克苏维埃
社会主义自治共和国
20 世纪 20 年代
长毛绒，丝线，丝绸

Skull cap

Karakalpaks. Karakalpak ASSR
1920s
Plush, silk thread, silk

139

丝绸面料

卡拉卡尔帕克人，德尔塔阿姆河地区 | 20 世纪初 | 丝绸

Silk fabric

Karakalpaks. Delta Amu Darya | Early 20th century | Silk

土库曼民族的丝绸织物
Sericulture of Turkmens

　　具有游牧文化的土库曼人从事蚕桑养殖和丝绸织造是欧亚丝绸史上的一个特殊现象。他们不使用专用织机，而是在一直以来用于生产羊毛织物的窄幅织机上织造丝绸。一种红色、边缘饰有黄色条纹且厚实的凯特尼布具有很高的地位，它一般用于缝制节日服装。这种布料是土库曼文化的一种象征，尽管从 20 世纪中叶开始，工厂使用不同颜色的粘胶丝或化纤线来制作服饰，但是到了 21 世纪，婚礼服饰的凯特尼布又用回了丝绸织造。

Silk farming and silk weaving among the Turkmens, a nation with a nomadic culture, is a unique phenomenon in the Eurasian silk history. They used to weave silk not with a designated machine, but with a narrow-beamed one, which was initially used for the production of wool fabrics. A dense heavy keteni fabric had high status, it usually was red with yellow stripes along the hems, and it was fabric meant for festive clothes. This fabric is one of the symbols of Turkmen culture, although from the mid-20th century, its production has moved from artisanal workshops to factories where it is made with viscous silk of different colors or synthetic fibers, in the 21st century the keteni for wedding garments is made with silk again.

女士套装
土库曼人
19 世纪末
Woman attire
Turkmens
Late 19th century

女士长袍斗篷

土库曼-特金人，外里海地区 | 19 世纪中叶 | 丝绸，羊毛织布，印花棉，丝线

Woman robe-like head cape

Turkmens-Teke. Transcaspian Oblast | Mid-19th century | Silk, woolen cloth, printed cotton, silk thread

女士长裙

土库曼人，土库曼苏维埃社会主义共和国

1930 ~ 1935 年

土库曼阿拉恰—丝绸，

印花布，棉织物

Woman dress

Turkmens. Turkmen SSR

1930-1935

Turkmen alacha-silk,

calico, cotton

女士长裤

土库曼人，外里海地区 | 19 世纪末 | 丝绸，棉织物，羊毛线，丝线

Woman trousers

Turkmens. Transcaspian Oblast | Late 19th century | Silk, cotton, wool thread, silk thread

女士头饰局部

土库曼-特金人，土库曼苏维埃社会主义共和国，
马里省和阿什哈巴德州
1925 ~ 1930 年
纸板，棉织物

Part of a woman headwear

Turkmens–Teke. Turkmen SSR,
Mary and Ashgabat Oblast
1925-1930
Cardboard, cotton

144

女士头巾

土库曼-特金人，土库曼苏维埃社会主义共和国，马里省和阿什哈巴德州 | 1925 ~ 1930 年 | 丝绸

Woman head kerchief

Turkmens–Teke. Turkmen SSR, Mary and Ashgabat Oblast | 1925-1930 | Silk

女士头巾

土库曼–特金人，土库曼苏维埃社会主
义共和国，马里省和阿什哈巴德州

1925 ～ 1930 年

丝绸

Woman head kerchief

Turkmens–Teke. Turkmen SSR, Mary and
Ashgabat Oblast

1925-1930

Silk

女士头巾

土库曼–特金人，土库曼苏维埃社会主义共和国，马里省和阿什哈巴德州 | 1925 ～ 1930 年 | 丝绸

Woman head kerchief

Turkmens–Teke. Turkmen SSR, Mary and Ashgabat Oblast | 1925-1930 | Silk

丝绸面料

土库曼人，土库曼苏维埃社会主义共和国

1930 ~ 1935 年

丝绸

Silk fabric

Turkmens. Turkmen SSR

1930-1935

Silk

146

丝绸面料

土库曼人，土库曼苏维埃社会主义共和国 | 1967 年 | 丝绸

Silk fabric

Turkmens. Turkmen SSR | 1967 | Silk

147

女士长袍斗篷
土库曼－约穆德人，里海东海岸
19 世纪下半叶
丝绸，羊毛织布，丝线
Woman robe-like head cape
Turkmens-Yomut. Eastern coast of the Caspian Sea
Second half of the 19th century
Silk, woolen cloth, silk thread

148

新娘面纱

土库曼–约穆德人，里海东海岸 | 19 世纪中叶 | 山东绸，丝线

Bride's face veil

Turkmens–Yomut. Eastern coast of the Caspian Sea | Mid-19th century | Shantung, silk thread

男式低帽

土库曼-乔多尔人，希瓦绿洲

1865 ～ 1875 年

丝绸，丝线，丝绵，植物纤维（棉絮）

Male underwear cap

Turkmens-Chowdur. Khiva oasis

1865–1875

Silk, silk thread, silk floss, plant fiber
(cotton wool)

男式低帽

土库曼-乔多尔人，希瓦绿洲

1865 ～ 1875 年

丝绸，棉织物，丝线，植物纤维（棉絮）

Male underwear cap

Turkmens-Chowdur. Khiva oasis

1865–1875

Silk, cotton, silk thread, plant fiber (cotton wool)

149

年轻女士头饰

土库曼-约穆德人，里海东海岸 | 19 世纪末 | 布，棉织物，羊毛，丝线

Young woman headwear

Turkmens-Yomut. Eastern coast of the Caspian sea | Late 19th century | Cloth, cotton, woolen, silk thread

4

伊朗文化中的
丝绸

Silk
in the Culture
of the Iran

丝线缠绕法
伊朗人，伊朗
1905年
Silk thread reeling
Iranians. Iran
1905

伊朗宫廷丝绸
Silk in Iran Court

伊朗是最早的丝绸生产中心之一。这里的气候——特别是在沿海省份——非常温和，有利于养蚕。当地的养蚕传统通过陆上和海上的丝绸贸易得到发展。伊朗国王沙赫是桑园和丝织生产的最大所有者，他还控制着丝绸出口，因丝绸贸易产生的高额关税充盈着国库。

国王的宫廷有专门的作坊，为王室和朝臣提供锦缎、天鹅绒和绸缎等昂贵的面料。这些作坊还制作华丽贵重的内饰织物，以此来展现国王宫殿的富丽堂皇。技艺高超的织工们从反面编织这些庞大的织物，借助镜子检查图案是否正确。

丝绒制作的宫廷地毯使伊朗的丝织业达到顶峰。地毯上的图饰以复杂但和谐的构图为妙，花纹极致精美。此外，这类丝毯是伊朗使节礼赠和精品出口贸易的重要组成部分。欧洲君主高度赞赏伊朗丝毯的丝绸光泽和精湛工艺，因此会购买丝毯制品用于装饰他们的"东方橱柜"和皇后的闺房。

Iran was one of the first centers of silk production. Here, silkworm breeding was facilitated by the climate, which was especially mild in the coastal provinces. The local sericulture traditions were enriched via the transit silk trade, both on land and at sea. The Iranian Shah was the largest owner of mulberry lands and silk-weaving production; he also controlled silk exports, and high duties on the silk trade replenished the state treasury.

Workshops functioned at the Shah's court, providing it with expensive fabrics: brocade, velvet and satin intended for royals' and courtiers' clothing. Those workshops also produced richly decorated, heavy interior fabrics designed to demonstrate the luxury of the Shah's palaces. Genius weavers processed huge canvases from the inside out, checking the pattern quality with the help of mirrors.

The pinnacle of Iranian silk weaving was the production of silk pile carpets. The carpets' ornament was characterized by an intricate yet harmonious composition, and its patterns were especially exquisite. In addition, these silk carpets were an important part of the embassy's gifts and Iran's elite export trade. European monarchs, who highly appreciated the silk glow and the subtlety of the work of Iranian carpet makers, purchased their products both for their 'oriental cabinets' and for the empresses' boudoirs.

祈祷毛绒毯

伊朗人，伊朗 | 19 世纪末 ~ 20 世纪初 | 丝绸，棉织物

Pile carpet

Iranians. Iran | Late 19th – early 20th century | Silk, cotton

154

祈祷毛绒毯

伊朗人，伊朗｜19 世纪末 ~ 20 世纪初｜丝绸，棉织物

Pile carpet

Iranians. Iran｜Late 19th – early 20th century｜Silk, cotton

伊朗城市文化中的丝绸
Silk in Iranian Urban Culture

　　伊朗丝织业的兴盛得益于多个省份大规模的丝绸生产，以及国内对丝绸和半丝绸织物的大量需求。在日常生活中，城市居民用丝织品制作沙发、床罩、枕头、新娘嫁妆专用餐巾、各种匣子和手提包。城市女性是丝织品的主要消费者，她们在市集上为自己和家人购买丝织品，包括上等的亚麻面料和密实的丝绸。

　　人们对特定类型丝绸的需求是由宫廷流行的时尚趋势决定的，随后传播到城市的宫廷圈层中。与粗棉布制成的低调街头服饰"卡多尔"形成鲜明对比的是伊朗富有女性穿着的昂贵而鲜艳的家居服饰，后者包括透明丝绸制成的短衬衫和绸缎或锦缎制成的上衣。

The heyday of silk weaving in Iran was determined by the large scale of silk production in a number of provinces as well as the high domestic demand for silk and semi-silk fabrics. In everyday life, townspeople used silk fabrics for sofa and bed covers, pillows, special napkins for the bride's dowry, various cases, and handbags. Townswomen were the main consumers of silk fabrics, buying them at city bazaars for themselves and their families, including fine linen fabrics and dense silks.

The demand for a particular type of silk was determined by fashion trends popular at the palace and then seeping through court circles in the urban. The chador, a discreet streetwear made of coarse cotton linen, was nothing like the expensive and vibrant domestic outfit of a rich Iranian woman. The latter included short shirts made of transparent silk fabric and upper jackets made of satin or brocade.

头巾

伊朗人，伊朗 ｜ 19 世纪下半叶 ｜ 丝绸，金属线

Head kerchief

Iranians. Iran ｜ Second half of the 19th century ｜ Silk, metal thread

女士外套

伊朗人，伊朗 ｜ 19 世纪下半叶 ｜ 缎，印花布

Woman jacket

Iranians. Iran ｜ Second half of the 19th century ｜ Satin, calico

女士外出套装
伊朗人
19 世纪末
Woman outing attire
Iranians
Late 19th century

穆斯林妇女的长面纱

伊朗人，伊朗
19 世纪下半叶
棉织物

The long veil of the Muslim woman

Iranians. Iran
Second half of the 19th century
Cotton

罩单

伊朗人，伊朗
19 世纪下半叶
棉织物，丝绸，玻璃珠

Covering sheet

Iranians. Iran
Second half of the 19th century
Cotton, silk, glass beads

161

女士长裤

伊朗人，伊朗 | 19 世纪下半叶 | 丝绸

Woman trousers

Iranians. Iran | Second half of the 19th century | Silk

织物

伊朗人，伊朗，卡尚市
20 世纪初
天鹅绒

Fabric

Iranians. Iran, Kashan
Early 20th century
Velvet

丝绸织物

伊朗人，伊朗，卡尚市 ｜ 19 世纪末 ｜ 缎

Silk fabric

Iranians. Iran, Kashan ｜ Late 19th century ｜ Satin

织物

伊朗人，伊朗，卡尚市
20 世纪初
天鹅绒

Fabric

Iranians. Iran, Kashan
Early 20th century
Velvet

163

织物

伊朗人，伊朗，卡尚市 | 20 世纪初 | 天鹅绒

Fabric

Iranians. Iran, Kashan | Early 20th century | Velvet

织物样品

伊朗人，伊朗，卡尚市 ｜ 20 世纪初 ｜ 丝绸

Fabric sample

Iranians. Iran, Kashan ｜ Early 20th century ｜ Silk

丝绸织物

伊朗人，伊朗，亚兹德市 | 20 世纪初 | 丝绸

Silk fabric

Iranians. Iran, Yazd | Early 20th century | Silk

5

高加索地区
居民文化中
的丝绸

Silk
in the Culture of
the Caucasus Regions

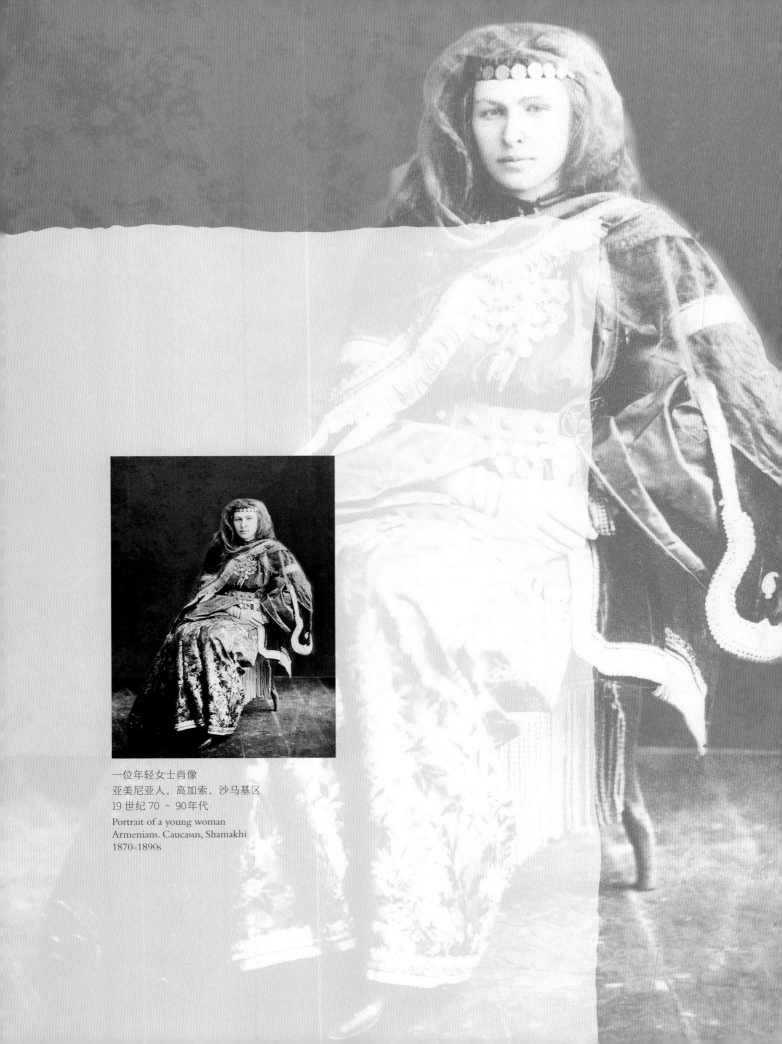

一位年轻女士肖像
亚美尼亚人，高加索，沙马基区
19 世纪 70 ~ 90 年代
Portrait of a young woman
Armenians. Caucasus, Shamakhi
1870–1890s

高加索东南部地区居民文化中的丝绸
Silk in the Culture of the South-Eastern Caucasus Population

包括库拉-阿拉克斯和里海沿岸低地在内的高加索东南部地区，养蚕业历史悠久并在 15 世纪依附于伊朗萨法维王朝的希尔凡沙赫时期达到了顶峰。

19 世纪，俄罗斯帝国将该地区纳入其版图，想尽一切办法促进该地区的丝绸生产发展。然而，当地居民还是以传统的方式从事养蚕业：女性培育蚕茧，男性缫丝。在沙马基、沙基、库巴和南高加索的其他城市，人们生产各种丝绸和半丝绸手工织物。一部分丝绸产品由当地居民购买，另一部分则通过批发贸易和旅行商人出口到高加索其他地区。其中，木板印花的"克拉加依"围巾至今仍是高加索东南地区非常有名的丝绸产品。

In the South-Eastern Caucasus Region, including the Kur-Araz and Caspian Lowlands, sericulture has been known since antiquity and reached its heyday in the 15th century in the country of Shirvanshahs under Safavid Iran.

In the 19th century, the region was joined by the Russian Empire, which in every possible way contributed to the development of silk production there. However, the local population engaged with sericulture in the old-fashioned way: women tended cocoons, and men did the reeling. In Shamakhi, Shaki, Quba, and other cities of the South Caucasus, artisans produced different types of silk and semi-silk fabrics. Some silk products circulated among the locals, others were exported for wholesale trade and picked up by traveling merchants to deliver them to other regions of the Caucasus. Among them, the "Kragai" headscarves is still a signature silk piece of the South-Eastern Caucasus.

北高加索和达吉斯坦地区民族文化中的丝绸
Silk in the Culture of the Peoples of the North Caucasus and Dagestan

19 世纪北高加索和达吉斯坦地区的养蚕史生动地说明了文化和政治因素对这一活动的制约性。19 世纪上半叶，哥萨克村庄的养蚕活动在国家的倡导下得到了发展，然而循序渐进、稳步发展的养蚕活动与北高加索山地居民的军事化生活习惯无法适配，20 世纪中叶开始，该地区的养蚕业日渐衰落。

进口丝绸弥补了当地丝绸产品的不足，但在山地社会严格标准化的文化下，进口丝绸在日常生活中无法占据一席之地。丝绸服装的主要受众为女性，由此来展现她们的家庭地位。比如，适婚年龄的女孩、新娘和新婚女性都应该穿着最华丽的丝绸服装。金线刺绣图案在北高加索女性服饰的装饰中意义非凡，女孩们还用金线刺绣装饰男士的钱包、烟袋和鼻烟盒，这些物品通常是节日和仪式的礼物。在达吉斯坦，女性的节日服装包括头饰、缎面刺绣长裤、精致的丝绸连衣裙，以及锦缎围裙。

The history of sericulture in the North Caucasus and Dagestan regions in the 19th century shows just how much it was influenced by culture and politics. In the first half of the 19th century, silkworm breeding in Cossack villages developed under state initiative, but the gradual and steady process of silkworm cultivation was not compatible with the militarized life of the North Caucasus population, since the mid-20th century, sericulture has been in decline.

The lack of local silk products was made up for with imports, and under the strict standardized culture of highland societies, they could not occupy a random place in everyday life. Silk clothes were mostly for women, it was a way to demonstrate their family status. The richest silk garbs were meant for girls of marriageable age, brides, and young married women. For example, the gold needlework played a special role in the decor of the North Caucasian women's costume. The girls also decorated men's purses, tobacco pouches, and snuffboxes with gold thread embroidery, these objects were often festive and ritual gifts. In Dagestan, women's festive attire included a headdress, satin embroidered trousers, a fine silk dress, as well as an apron of rich brocade.

阿塞拜疆人
Azerbaijani

阿塞拜疆人是主要生活在今阿塞拜疆共和国和伊朗西北部阿塞拜疆地区的突厥人，主要由定居的农民和牧民组成。大多数阿塞拜疆人是什叶派穆斯林。他们讲阿塞拜疆语，属于突厥语西南分支。

阿塞拜疆人的民族来源复杂，其最古老的血统是来自外高加索东部的土著居民，也可能是来自波斯北部的梅迪亚人。他们在伊朗萨珊王朝时期（公元 3 ~ 7 世纪）逐渐被波斯化。当地人群的突厥化可以追溯到 11 世纪塞尔柱突厥人对该地区的征服，以及随后几个世纪突厥人的不断涌入，包括 13 世纪蒙古人征服期间突厥人的迁入。

Azerbaijani, is any member of a Turkic people living chiefly in the Republic of Azerbaijan and in the region of Azerbaijan in northwestern Iran. They are mainly sedentary farmers and herders. Most Azerbaijani are Shia Muslims. They speak Azerbaijani, a language belonging to the southwestern branch of Turkic languages.
The Azerbaijani are of mixed ethnic origin, the oldest element deriving from the indigenous population of eastern Transcaucasia and possibly from the Medians of northern Persia. This population was Persianized during the period of the Sassanid period in Iran (3rd-7th century). Turkicization of the population can be dated from the region's conquest by the Seljuq Turks in the 11th century and the continued influx of Turkic populations in subsequent centuries, including those groups that migrated during the Mongol conquests in the 13th century.

阿瓦尔人
Avars

阿瓦尔人是高加索东北部的民族。阿瓦尔人居住在黑海和里海之间的北高加索地区。与北高加索地区的其他民族一样，阿瓦尔人居住在海拔约 2000 米的古老村落中。高加索阿瓦尔人使用的阿瓦尔语属于东北高加索语系。自 14 世纪以来，阿瓦尔人主要信仰逊尼派伊斯兰教。

169

The Avars are a Northeast Caucasian ethnic group. The Avars reside in the North Caucasus between the Black Sea and the Caspian Sea. Alongside other ethnic groups in the North Caucasus region, the Avars live in ancient villages located approximately 2,000 meters above sea level. The Avar language spoken by the Caucasian Avars belongs to the family of Northeast Caucasian languages. Sunni Islam has been the prevailing religion of the Avars since the 14th century.

女士套装
阿塞拜疆人
20 世纪初
Woman attire
Azerbaijanis
Early 20th century

头巾
阿塞拜疆人
20 世纪初
丝绸

Head kerchief
Azerbaijanis
Early 20th century
Silk

女士外套

阿塞拜疆人
20 世纪初
丝绸，印花布，金线，棉絮

Woman jacket

Azerbaijanis
Early 20th century
Silk, calico, gold thread, cotton wadding

带绳短裙

阿塞拜疆人
20 世纪初
棉织物，印花布，金线，丝线

Corded skirt

Azerbaijanis
Early 20th century
Cotton, calico, gold thread, silk thread

172

头饰

阿塞拜疆人
20 世纪初
丝绸，金线

Headwear

Azerbaijanis
Early 20th century
Silk, gold thread

173

女士衬衫

阿塞拜疆人，高加索，伊斯梅林地区，阿塞拜疆苏维埃社会主义共和国 | 20 世纪初 | 棉织物，丝绸，丝线，金属线，珍珠母贝

Woman shirt

Azerbaijanis. Caucasus, Ismayilli District, Azerbaijan SSR | Early 20th century | Cotton, silk, silk thread, metal thread, mother-of-pearl

马鞍装饰

阿塞拜疆人，高加索 | 20 世纪初 | 丝绸，棉织物，羊毛，皮革

Saddle cover

Azerbaijanis. Caucasus | Early 20th century | Silk, cotton, wool, leather

头巾

阿塞拜疆人，高加索 ｜ 20 世纪初 ｜ 丝绸

Head kerchief

Azerbaijanis. Caucasus ｜ Early 20th century ｜ Silk

女士套装
阿瓦尔人
20 世纪初
Woman attire
Avars
Early 20th century

围裙

阿瓦尔人，高加索

20 世纪初

丝绸，棉织物，金属线

Apron

Avars. Caucasus

Early 20th century

Silk, cotton, metal thread

180

女士外套

阿瓦尔人，列兹金 | 19 世纪下半叶 | 丝绸

Woman outerwear

Avars. Lezgins | Second half of the 19th century | Silk

面纱

阿瓦尔人，高加索

20 世纪初

丝绸，金属线

Veil

Avars. Caucasus

Early 20th century

Silk, metal thread

女士头饰

阿瓦尔人，高加索｜20 世纪初｜丝绸，印花布，金属线（拉丝银）

Woman headwear

Avars. Caucasus｜Early 20th century｜Silk, calico, metal thread (silver wire)

182

女士衬衫

阿瓦尔人，高加索 | 20 世纪初 | 丝绸，金属线

Woman shirt

Avars. Caucasus | Early 20th century | Silk, metal thread

女士长裤

阿瓦尔人，高加索 | 20 世纪初 | 缎，印花布，金属线

Woman trousers

Avars. Caucasus | Early 20th century | Satin, calico, metal thread

枕头刺绣样品

达尔金人，高加索，达吉斯坦地区，
凯塔戈—塔巴萨兰区
19 世纪
棉织物，丝线

Embroidery sample for the pillow case

Dargins. Caucasus, Dagestan Oblast,
Kaytago-Tabasaranskiy okrug
19th century
Cotton, silk thread

184

窗帘

库巴钦人，高加索 ｜ 19 世纪末 ~ 20 世纪初 ｜ 锦缎，金属线

Curtain

Kubachi. Caucasus ｜ Late 19th - early 20th century ｜ Brocade, metal thread

女士套装
卡拉恰耶夫人
20 世纪初
Woman attire
Karachays
Early 20th century

女士头饰

卡拉恰耶夫人，高加索，卡拉恰耶区
19 世纪末
天鹅绒，金属线

Woman headwear

Karachays. Caucasus, Karachay
Late 19th century
Velvet, metal thread

扣环

巴尔卡尔人
19 世纪末 ~ 20 世纪初
白色金属，玻璃

Ring fastener

Balkarians
Late 19th – early 20th century
White metal, glass

女士长裙

卡拉恰耶夫人 | 20 世纪初 | 凸纹布，丝绸，金银饰带

Woman dress

Karachays | Early 20th century | Cameo, silk, passementerie

卡巴尔德人
20 世纪初
丝线

Shawl

Kabardians
Early 20th century
Silk thread

衬衫
卡拉恰耶夫人
20 世纪初
丝绸，棉线

Shirt

Karachays
Early 20th century
Silk, cotton thread

187

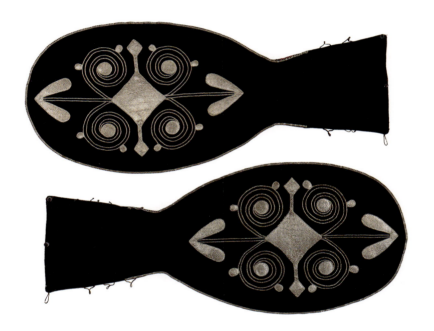

女士长裙袖子垂坠

卡巴尔德人，巴尔卡尔，高加索 | 20 世纪初 | 丝绸，羊毛织布，金属线

Sleeve decoration for a woman dress

Kabardians. Balkars, Caucasus | Early 20th century | Silk, woolen cloth, metal thread

高加索西南地区民族文化中的丝绸：拜占庭遗产
Silk in the Ethnoculture of the Southwestern Caucasus: the Byzantine Heritage

　　高加索西南部黑海地区（包括历史上的亚美尼亚和格鲁吉亚区域）的丝绸生产传统与拜占庭遗产密不可分，该丝绸生产地区可被视为小亚细亚地区的一部分。这些织物上的图案在现代早期受到欧洲影响：巴洛克风格的法式玫瑰成为主要的装饰图案之一。

　　卡尔斯、巴图姆、梯弗里斯、库泰斯、阿哈尔捷克及其他主要城市是各种商品的流通地，这些商品包括丝绸原料和丝织品，这促进了丝绸服装在城镇居民中的广泛传播。丝绸刺绣便是一种常见的女性手工艺，主要用于装饰女性服饰的配件，如胸衣、腰带等。

The traditions of silk production in the Black Sea area of the Southwestern Caucasus, which includes the historical lands of Armenia and Georgia, are inextricably intertwined with the heritage of Byzantium, the aforementioned sericulture province can be considered a part of Asia Minor. These brocade patterns were under European influence during the Early Modern Period – the baroque 'French rose' became one of the leading ornamental motifs.

Kars, Batumi, Tiflis, Kutaisi, Akhaltsikhe, and other major cities had a heavy trade traffic of various goods, including silk raw materials and silk fabrics, which contributed to the distribution of silk garments among the residents. For example, silk embroidery was a common type of women's handiwork that was supposed to decorate primarily accessories and additional elements of women's costumes, like breastplates, belts, etc.

亚美尼亚人
Armenians

　　亚美尼亚人，这一拥有古老文化的族群，最初生活在被称为亚美尼亚的地区，即今天的土耳其东北部和亚美尼亚共和国。虽然有些亚美尼亚人仍生活在土耳其，但亦有 300 多万亚美尼亚人生活在亚美尼亚共和国；还有大量亚美尼亚人生活在格鲁吉亚、高加索和中东等其他地区。直到 20 世纪 80 年代末，还有大量亚美尼亚人居住在阿塞拜疆。

Armenians, an ethnic group with an ancient culture, originally lived in the area known as Armenia, which comprised what is now northeastern Turkey and the Republic of Armenia. Although some remain in Turkey, more than three million Armenians live in the republic; large numbers also live in Georgia as well as other areas of the Caucasus and the Middle East. A large number lived in Azerbaijan until the late 1980s.

挂钟外壳

切尔克斯人，高加索 | 20 世纪初 | 天鹅绒，印花布，金属线，玻璃

Clock case

Circassians. Caucasus | Early 20th century | Velvet, calico, metal thread, glass

面纱

奥塞梯人，高加索，新奥塞廷斯卡亚哥萨克村 ｜ 20 世纪中叶 ｜ 丝绸

Veil

Ossetians. Caucasus, Novo-Osetinskaya Cossack village (stanitsa) ｜ Mid-20th century ｜ Silk

男式外套

斯塔夫罗波尔土库曼人，高加索，斯塔夫罗波尔省 | 20 世纪初 | 丝绸，锦缎

Male outerwear

Stavropol Turkmen. Caucasus, Stavropol Province | Early 20th century | Silk, brocade

女士套装
亚美尼亚人
19 世纪末
Woman attire
Armenians
Late 19th century

眉带（头饰的一部分）

亚美尼亚人

20 世纪上半叶

天鹅绒，织物，线

Browband (part of headwear)

Armenians

First half of the 20th century

Velvet, fabric, thread

额饰

亚美尼亚人

19 世纪末

白色金属，黄色金属，织物

Forehead Jewellery

Armenians

Late 19th century

White metal, yellow metal , fabric

194

头饰

亚美尼亚人 | 19 世纪末 | 玻璃珠，铜合金，织物

Head ornament

Armenians | Late 19th century | Glass beads, copper alloy, fabric

项链

亚美尼亚人，第比利斯省，阿哈尔齐赫地区

19 世纪末

青铜，玻璃，绦

Necklace

Armenians. Tiflis Province, Akhaltsikhe Uyezd

Late 19th century

Copper, glass, narrow braid silk

项链

亚美尼亚人

19 世纪末

青铜

Necklace

Armenians

Late 19th century

Bronze

195

女士衣带

亚美尼亚人，第比利斯省，阿哈尔齐赫地区 ｜ 19 世纪末 ｜ 丝绸

Woman sash

Armenians. Tiflis Province, Akhaltsikhe Uyezd ｜ Late 19th century ｜ Silk

女士头巾

亚美尼亚人

20 世纪中叶

半丝织物

Woman kerchief

Armenians

Mid-20th century

Semi-silk

女士长裙

亚美尼亚人

19 世纪末 ~ 20 世纪初

丝绸，棉织物，金线

Woman dress

Armenians

Late 19th – early 20th century

Silk, cotton, gold thread

196

女士长裙

亚美尼亚人 | 20 世纪初 | 天鹅绒，白色金属，棉织物，金线

Woman dress

Armenians | Early 20th century | Velvet, white metal, cotton, gold thread

披巾

亚美尼亚人

19 世纪末

缎，印花布

Scarf

Armenians

Late 19th century

Satin, calico

围裙

亚美尼亚人，高加索，巴统省 | 19 世纪末 ~ 20 世纪初 | 棉织物，天鹅绒，羊毛织布，金属线

Apron

Armenians. Caucasus, Batum Oblast | Late 19th – early 20th century | Cotton, velvet, woolen cloth, metal thread

女士头饰

亚美尼亚人，第比利斯省，阿哈尔齐赫地区
19 世纪末
羊毛织布，丝绸，金属线

Woman headwear

Armenians. Tiflis Province, Akhaltsikhe Uyezd
Late 19th century
Woolen cloth, silk, metal thread

头饰局部

亚美尼亚人 | 19 世纪末 ~ 20 世纪初 | 棉织物，金线

Part of the headwear

Armenians | Late 19th – early 20th century | Cotton, gold thread

女士长裙

亚美尼亚人，第比利斯省，阿哈尔齐赫地区 ｜ 19 世纪末 ~ 20 世纪初 ｜ 印花布，锦缎，丝线，金属线

Woman dress

Armenians. Tiflis Province, Akhaltsikhe Uyezd ｜ Late 19th – early 20th century ｜ Calico, brocade, silk thread, metal thread

女士家居衬衫

亚美尼亚人，第比利斯省，阿哈尔齐赫地区 | 19 世纪末 ~ 20 世纪初 | 棉织物，丝线，金属线，金属（亮片）

Woman bath shirt

Armenians. Tiflis Province, Akhaltsikhe Uyezd | Late 19th – early 20th century | Cotton, silk thread, metal thread, metal (sequin)

刺绣样品

亚美尼亚人，高加索 | 19 世纪末 ~ 20 世纪初 | 亚麻布

Embroidery sample

Armenians. Caucasus | Late 19th – early 20th century | Linen

刺绣样品

亚美尼亚人，高加索 | 19 世纪末 ~ 20 世纪初 | 亚麻布，丝绸

Embroidery sample

Armenians. Caucasus | Late 19th – early 20th century | Linen, silk

刺绣样品

格鲁吉亚人，高加索 | 20 世纪初 | 亚麻布，丝绸

Embroidery sample

Georgians. Caucasus | Early 20th century | Linen, silk

女士上衣

格鲁吉亚人，高加索 │ 19 世纪中叶 │ 丝绸，棉织物，铜

Woman outerwear

Georgians. Caucasus │ Mid-19th century │ Silk, cotton, copper

女士刺绣头饰

格鲁吉亚人，高加索

20 世纪初

天鹅绒

Embroidery for a woman headwear

Georgians. Caucasus

Early 20th century

Velvet

女士刺绣头饰

格鲁吉亚—卡尔塔林人，高加索

20 世纪初

天鹅绒，丝绸

Embroidery for a woman headwear

Georgians – Kartli. Caucasus

Early 20th century

Velvet, silk

餐巾

格鲁吉亚人，高加索 | 19 世纪末 | 丝绸，金属线

Serviette

Georgians. Caucasus | Late 19th century | Silk, metal thread

长裙前部装饰

格鲁吉亚人，高加索 ｜ 20 世纪初 ｜ 丝绸，金属线

Decorated front part of a dress

Georgians. Caucasus ｜ Early 20th century ｜ Silk, metal thread

高加索地区
居民文化中的丝绸

209

女士长裙腰带

格鲁吉亚人，高加索 | 20 世纪初 | 丝绸，金属线

Sash for a woman dress

Georgians. Caucasus | Early 20th century | Silk, metal threa

钱袋

格鲁吉亚人，高加索

20 世纪初

丝线，金属线，骨头

Purse

Georgians. Caucasus

Early 20th century

Silk, metal thread, bone

210

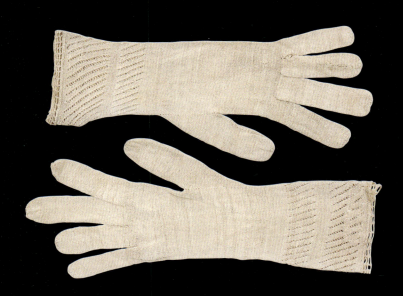

手套

格鲁吉亚人，高加索 │ 20 世纪初 │ 丝绸

Gloves

Georgians. Caucasus │ Early 20th century │ Silk

蕾丝带

格鲁吉亚—卡赫提亚人，高加索

20 世纪初

丝绸

Lace

Georgians - Kakhetians. Caucasus

Early 20th century

Silk

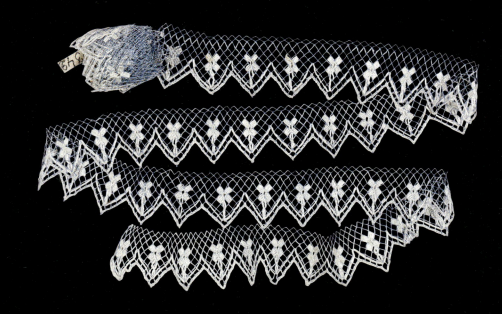

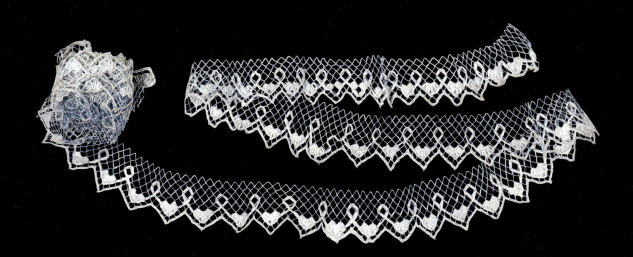

蕾丝带

格鲁吉亚—卡赫提亚人，高加索 ｜ 20 世纪初 ｜ 丝绸

Lace

Georgians - Kakhetians. Caucasus ｜ Early 20th century ｜ Silk

6

鞑靼
和巴什基尔地区
文化中的丝绸

Silk
in the Culture of
Tatar and Bashkir Regions

家庭合影
巴什基尔人，东欧，南乌拉尔
1914年
Family
Bashkirs. Eastern Europe, Southern Urals
1914

女人和女孩在刺绣
马里人，马里埃尔苏维埃社会主义自治共和国
1939年
Woman and girl doing embroidery
Mari. Mari ASSR
1939

早在中世纪，各种丝织品就被鞑靼人和巴什基尔人所熟知。丝绸是一种昂贵的商品，通过中东和中亚的贸易路线传入该地区。无论对于城市还是农村的女性都通过穿着丝绸服装来彰显她们富有的社会地位。18世纪到20世纪初，富有的鞑靼女性可以用锦缎或薄绸缝制整套服装。而丝线则在普通民众中特别流行，人们使用丝线缝制衣服、制作室内装饰——毛巾、窗帘和祈祷毯；用丝绸缎带、织带和金银饰带装饰胸巾、肩带、裙边、头饰和上衣。

19世纪中叶，手工制作的中亚半丝绸面料在鞑靼人中开始流行，主要是条纹贝卡萨布织物和阿布尔织物。条纹织物主要用于制作男士长衫，而阿布尔织物则用于制作裤子、上衣以及装饰胸衣和女帽。20世纪初，由于鞑靼和巴什基尔传统文化的普遍城市化，这种流行趋势消失了。

19世纪下半叶，随着俄罗斯纺织业的发展，丝绸材料在居民中的普及程度达到顶峰。此时，绸缎和天鹅绒成为缝制无袖背心最常见的面料，男女头饰——即无边便帽和卡尔法克（鞑靼族女士帽子）也主要使用天鹅绒缝制。当时，丝绸披肩和机织镂空披肩开始流行。比如，到20世纪初，女性节日服装中最流行的元素就是用丝线编织的镂空披肩。

The Tatars and Bashkirs have been familiar with a variety of silk fabrics since the Middle Ages. It was an expensive commodity that was imported into the region via trade routes crossing the Middle East and Central Asia. Both urban and rural women demonstrated their wealthy social status by wearing silk clothes. In the 18th century and early 20th century, wealthy Tatar women could have entire dresses sewn with brocade or gauze. Silk thread was especially popular with the general population, being used for the embroidery of clothes and interior decor fabrics such as towels, curtains, and prayer carpets. Breastplates, sashes, hems of women's dresses, headgear, and outerwear were decorated with silk ribbons and webbing, as well as passementerie.

In the middle of the 19th century, handmade Central Asian semi-silk fabrics became popular among the Tatars, mainly striped bekasab or abr fabrics. Striped fabrics were mainly used for production of men's robes, trousers and outerwear were sewn with abr fabrics that were also used in the decoration of breastplates and women's hats, too. In the early 20th century, this trend disappeared due to the general urbanization of the traditional culture of Tatars and Bashkirs.

The accessibility of silk materials peaked in the second half of the 19th century with the development of the textile industry in Russia. At this time, satin and velvet became the most common fabrics for sewing sleeveless vests, men's and women's headwear, like tubeteika and kalfaks, were also made mainly of velvet. Silk and factory-produced shawls became quite common. For example, by the early 20th century, the most popular element of festive women's attire was an openwork shawl made of silk warp-knitted thread.

鞑靼人
Tatars

鞑靼人是东欧和亚洲各地以"鞑靼"为名的突厥族群的总称。最初，"鞑靼"这个民族名称可能是指鞑靼部落。成吉思汗统一草原各部落后，该部落联盟最终并入蒙古帝国。历史上，鞑靼人一词被用来指代来自北亚和中亚大陆的被称作"鞑靼"地区的人。最近，这个词被狭义地用来代表那些自称为鞑靼人或使用鞑靼语的族群。

迄今为止，鞑靼人中规模最大的群体是伏尔加鞑靼人，他们是俄罗斯欧洲区域伏尔加—乌拉尔地区的原住民，占鞑靼斯坦人口的53%。他们使用的语言被称为鞑靼语。截至2010年，俄罗斯约有530万鞑靼族人。沙皇俄国和俄罗斯帝国的许多贵族家庭成员都有鞑靼血统。

The Tatars is an umbrella term for different Turkic ethnic groups bearing the name "Tatar" across Eastern Europe and Asia. Initially, the ethnonym Tatar possibly referred to the Tatar confederation. That confederation was eventually incorporated into the Mongol Empire when Genghis Khan unified the various steppe tribes. Historically, the term Tatars (or Tartars) was applied to anyone originating from the vast Northern and Central Asian landmass then known as Tartary. More recently, the term has come to refer more narrowly to related ethnic groups who refer to themselves as Tatars or who speak languages that are commonly referred to as Tatar.

The largest group amongst the Tatars by far are the Volga Tatars, native to the Volga-Ural region of European Russia. They compose 53% of the population in Tatarstan. Their language is known as the Tatar language. As of 2010, there were an estimated 5.3 million ethnic Tatars in Russia. Many noble families in the Tsardom of Russia and the Russian Empire had Tatar origins.

鞑靼皇后 Söyembikä 肖像
Portrait of Söyembikä, Empress of Tartary

巴什基尔人
Bashkirs

巴什基尔人是俄罗斯的基普恰克-布尔加突厥族。他们主要聚居在俄罗斯联邦的巴什科尔托斯坦共和国和巴日加德这一历史悠久的地区，该地区横跨乌拉尔山脉两侧，是东欧和北亚的交汇处。鞑靼斯坦共和国、彼尔姆边疆区、车里雅宾斯克州、奥伦堡州、秋明州、斯维尔德洛夫斯克州、库尔干州和俄罗斯其他地区也存在较小规模的巴什基尔人社区，哈萨克斯坦和乌兹别克斯坦也有相当数量的巴什基尔人。

大多数巴什基尔人讲巴什基尔语，与属于突厥语基普恰克语支的鞑靼语和哈萨克语关系密切。他们与突厥民族联系密切。巴什基尔人以前是游牧民族，16 世纪起被俄罗斯统治。此后，他们在俄罗斯历史上发挥了重要作用，最终在俄罗斯帝国、苏联和后苏联时期获得了自治地位。

The Bashkirs are a Kipchak-Bulgar Turkic ethnic group indigenous to Russia. They are concentrated in Bashkortostan, a republic of the Russian Federation and in the broader historical region of Badzhgard, which spans both sides of the Ural Mountains, where Eastern Europe meets North Asia. Smaller communities of Bashkirs also live in the Republic of Tatarstan, the oblasts of Perm Krai, Chelyabinsk, Orenburg, Tyumen, Sverdlovsk and Kurgan and other regions in Russia, sizable minorities exist in Kazakhstan and Uzbekistan.

Most Bashkirs speak the Bashkir language, closely related to the Tatar and Kazakh languages, which belong to the Kipchak branch of the Turkic languages, they share historical and cultural affinities with the broader Turkic peoples. Previously nomadic and fiercely independent, the Bashkirs gradually came under Russian rule beginning in the 16th century. They have since played a major role throughout the history of Russia, culminating in their autonomous status within the Russian Empire, Soviet Union and post-Soviet Russia.

215

芬兰乌戈尔和楚瓦什地区文化中的丝绸
Silk in the Cultures of the Finno-Ugric and Chuvash Regions

　　远古时代直到 20 世纪 30 年代，伏尔加河地区和乌拉尔地区的楚瓦什人和芬兰乌戈尔人普遍穿着自制织物制作的衣服。人们用白色家纺粗布缝制衬衫、长衫、手帕、女式胸衣、腰带坠饰和头饰，并用平针和图案刺绣装饰。丝线在刺绣中通常起着主要作用，丝线的光泽和闪烁的色彩为刺绣构图的艺术设计奠定了基础。许多衣服通常还会用丝线缝上流苏或褶边。人们一直使用天然染料为丝绸染色，自 19 世纪末才开始使用人工染料。

　　节日盛装的接缝处和下部边缘通常用丝绸花边和丝带装饰，它们可以在家中染色，也可以从集市和行商小贩那里买来现成的进行装饰。这些装饰还用于点缀胸衣、女士头饰的前额带和外衣镶边。19 世纪末尤为流行用镶边装饰衣服，人们使用棋盘格花粗布和机织制品代替白色粗麻布。

　　19 世纪下半叶，丝绸围巾和机织镂空披肩在伏尔加地区广为流行。

　　科米—伊泽米亚人通常在女性节日服装中使用丝织品。19 世纪至 20 世纪初，节日服装包括一件衬衫和一件"萨拉凡"（传统民族服装，一种无袖长裙），其大部分由俄罗斯工厂生产的丝织品制成。

From ancient times until the 1930s, the Chuvash and Finno-Ugric peoples of the Volga region and the Urals wore clothes made of homemade fabrics. Shirts, robes, kerchiefs, women's breastplates, waist pendants, and headwear made of white artisanal canvas were decorated with counted-thread and double-running stitch embroidery. It focused mostly on silk threads, the shimmer and vibrant colors of which provided the basis for the artistic solutions of embroidered compositions. Many clothing pieces were often additionally decorated at the edges with tassels or fringes, as well as silk threads. Natural dyes were used in silk dyeing until the late 19th century when they gave way to artificial dyes.

Elements of the festive costumes were often decorated at the seams and lower edges with silk webbing and ribbons, dyed at home or ready-bought at fairs and traveling merchants. They were used mainly in the manufacture of breastplates, forehead bands for women's headgear, and outerwear trimming. Trimming became especially popular in the late 19th century when the white canvas was replaced by a checkered buntgewebe and factory fabrics.

In the second half of the 19th century, silk scarves and factory-produced shawls became common in the Volga region.

Izhma Komi most proactively used silk fabrics for women's festive attire. In the 19th and early 20th centuries, it was traditionally comprised of a shirt and a sarafan (a traditional national costume, a sleeveless dress), partially or completely made of purchased silk fabrics produced by Russian factories.

男式套装
喀山鞑靼人
19 世纪下半叶
Male attire
Kazan Tatars
Second half of the 19th century

男式无边便帽

喀山鞑靼人，东欧
19 世纪中叶
天鹅绒，半丝织物（内衬），金属线，
棉线，黄色合金（小亮片）

Male skull-cap

Kazan Tatars. Eastern Europe
Mid-19th century
Velvet, semi-silk (lining), metal thread,
cotton thread, yellow alloy (sequins)

男式长袍

喀山鞑靼人，东欧，喀山地区
19 世纪下半叶
半丝织物，印花布（内衬），棉织物

Male robe

Kazan Tatars. Eastern Europe, Kazan Province
Second half of the 19th century
Semi-silk, calico (lining), cotton

男式衬衫

鞑靼人，中伏尔加鞑靼人，米舍尔亚克人
19 世纪末 ~ 20 世纪初
印花布

Male shirt

Tatars. Central Volga Tatars, Mischeryaks
Late 19th – early 20th century
calico

男式无边小帽

喀山鞑靼人，东欧
19 世纪末 ~ 20 世纪初
天鹅绒，棉织物，金属线

Male skull-cap

Kazan Tatars. Eastern Europe
Late 19th – early 20th century
Velvet, cotton, metal thread

220

男式长裤

鞑靼人，东欧，喀山地区 | 19 世纪末 | 半丝织物，棉织物

Male trousers

Tatars. Eastern Europe, Kazan Province | Late 19th century | Semi-silk, cotton

帽子
鞑靼人
19 世纪末 ~ 20 世纪初
天鹅绒，珍珠，棉织物，纸板

Cap
Tatars
Late 19th – early 20th century
Velvet, pearls, cotton, cardboard

女士长裙（复制品）
鞑靼人，伏尔加河中游，喀山地区 │ 2019 年 │ 丝绸，棉织物

Woman dress (replica)
Tatars. Middle Volga, Kazan Province │ 2019 │ Silk, cotton

221

女士套装
喀山鞑靼人
20 世纪初
Woman attire
Kazan Tatars
Early 20th century

女士无袖外套

喀山鞑靼人，东欧，喀山地区
20 世纪初
天鹅绒，棉织物（内衬），金银饰带，
棉线

Woman sleeveless jacket

Kazan Tatars. Eastern Europe, Kazan
Province
Early 20th century
Velvet, cotton (lining), passementerie,
cotton thread

喀山鞑靼人，东欧，喀山地区
20 世纪初
丝绸

Veil

Kazan Tatars. Eastern Europe, Kazan Province
Early 20th century
Silk

少女帽

喀山鞑靼人，东欧，喀山地区
19 世纪末 ~ 20 世纪初
天鹅绒，棉织物（内衬），金属线，纸板，
黄色合金（小亮片）

Girl's cap

Kazan Tatars. Eastern Europe, Kazan Province
Late 19th – early 20th century
Velvet, cotton (lining), metal thread, cardboard,
yellow alloy (sequins)

女士头饰

鞑靼人，东欧
19 世纪末
天鹅绒，金属线，纸板，黄色合金（小亮片）

Woman headwear

Tatars. Eastern Europe
Late 19th century
Velvet, metal thread, cardboard, yellow alloy (sequins)

女士头饰

鞑靼人，东欧 | 19 世纪末 | 天鹅绒，金属线，纸板，黄色合金（小亮片）

Woman headwear

Tatars. Eastern Europe | Late 19th century | Velvet, metal thread, cardboard, yellow alloy (sequins)

女士头饰

喀山鞑靼人，东欧，喀山地区

19 世纪中叶

天鹅绒，锦缎，棉织物（内衬），丝绸，金属线，
玻璃（小珠子），白色合金（小亮片）

Woman headwear

Kazan Tatars. Eastern Europe, Kazan Province

Mid-19th century

Velvet, brocade, cotton (lining), silk, metal thread,
glass (beads), white alloy (sequins)

额带

巴什基尔人，东欧，奥伦堡地区，上乌拉尔地区

19 世纪

布，丝绸，羊毛

Forehead band

Bashkirs. Eastern Europe, Orenburg Province, Verkhneuralsky Uyezd

19th century

Cloth, silk, wool

披肩

鞑靼人—克拉珊人，东欧，维亚特卡地区，叶拉布加地区 ｜ 19 世纪 ｜ 布，丝绸

Kerchief

Tatars – Kryashens. Eastern Europe, Vyatka Province, Yelabuzhsky Uyezd ｜ 19th century ｜ Cloth, silk

新郎头巾

楚瓦什人，东欧，乌法地区，别列别伊地区 ｜ 19 世纪 ｜ 布，丝带，丝绸，羊毛

Groom's kerchief

Chuvash. Eastern Europe, Ufa Province, Belebeyevsky Uyezd ｜ 19th century ｜ Cloth, silk (ribbons), silk, wool

餐巾

巴什基尔人，东欧，奥伦堡地区，特洛伊茨克地区 | 19 世纪 | 布，丝绸

Napkin

Bashkirs. Eastern Europe, Orenburg Province, Troitsky Uyezd | 19th century | Cloth, silk

毛巾

巴什基尔人，东欧，奥伦堡地区，
车里雅宾斯克地区
19 世纪
布，丝绸

Towel

Bashkirs. Eastern Europe, Orenburg
Province, Chelyabinsk Uyezd
19th century
Cloth, silk

毛巾尾部

巴什基尔人 ┃ 19 世纪末 ~ 20 世纪初 ┃ 布，丝线，丝绸，饰带

Towel end

Bashkirs ┃ Late 19th - early 20th century ┃ Cloth, silk thread, silk, sash

女士节日套装
乌德穆尔特人
19 世纪末 - 20 世纪初
Woman festive attire
Udmurts
Late 19th - early 20th century

女士头饰

乌德穆尔特人
20 世纪初
亚麻布，粗花布，印花布，金
银织带，蕾丝

Woman Headwear

Udmurts
Early 20th century
Linen, coarse flower cloth, calico,
gold and silver tapestry, lace

头巾

乌德穆尔特人，东欧
19 世纪末 ~ 20 世纪初
布，丝绸

Headband towel

Udmurts. Eastern Europe
Late 19th – early 20th century
Cloth, silk

女士头罩

摩尔多瓦人，东欧，辛比尔斯克地区，阿尔达托夫地区 | 19 世纪末 ~ 20 世纪初 | 丝绸

Woman head cover

Mordvins Erzya. Eastern Europe, Simbirsk Province, Ardatovsky Uyezd | Late 19th – early 20th century | Silk

女士头巾

摩尔多瓦人，东欧，
辛比尔斯克地区，
阿尔达托夫地区
19 世纪末 ~ 20 世纪初
丝绸，丝线

Woman kerchief

Mordvins Erzya.
Eastern Europe,
Simbirsk Province,
Ardatovsky Uyezd
Late 19th – early 20th century
Silk, silk thread

女士披巾

乌德穆尔特人
19 世纪末
布，丝绸

Woman scarf

Udmurts,
Late 19th century
Cloth, silk

233

围裙

乌德穆尔特人，东欧 | 19 世纪末 ~ 20 世纪初 | 布，丝带，棉织物，金属线，纱线和羊毛线，绦带，白色合金

Apron

Udmurts. Eastern Europe | Late 19th– early 20th century | Cloth, silk (ribbons), cotton, metal thread, lic and wool thread, narrow braid, white alloy

项链

乌德穆尔特人，东欧
20 世纪初
玻璃珠，棉织物，铜（坠饰），
贝壳

Necklace

Udmurts. Eastern Europe
Early 20th century
Glass (beads), cotton, copper
(pendants), shells

女士颈饰

乌德穆尔特人，东欧，乌德穆尔特
苏维埃社会主义自治共和国，基洛
夫地区，格拉佐夫县
20 世纪 30 年代
玻璃珠，黄色合金（坠饰），棉织物，
贝壳

Woman neck decoration

Udmurts. Eastern Europe, Udmurts
ASSR, Kirov Oblast, Glazovsky Uyezd
1930s
Glass (beads), yellow alloy (pendants),
cotton, shells

女士装饰肩带

乌德穆尔特人，东欧
20 世纪初
小喇叭，棉织物，黄色合金，贝壳

Woman decorative shoulder strap

Udmurts. Eastern Europe
Early 20th century
Bugle, cotton, yellow alloy, shells

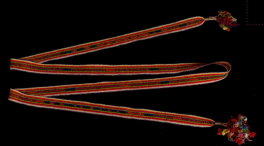

234

腰带

乌德穆尔特人 │ 20 世纪初 │ 棉线，羊毛线

Belt

Udmurts │ Early 20th century │ Cotton thread, wool thread

女士衬衫

乌德穆尔特人，东欧
20 世纪 20 年代
布，丝带，丝绸，羊毛

Woman shirt

Udmurts. Eastern Europe
1920s
Cloth, silk (ribbons), silk, wool

235

女士外套（卡夫坦长衫）

乌德穆尔特人，东欧 ｜ 20 世纪 30 年代 ｜ 布，丝带，棉织物，羊毛，棉线，丝绸，白色合金

Woman outerwear – kaftan

Udmurts. Eastern Europe ｜ 1930s ｜ Cloth, silk (ribbons), cotton, wool, cotton thread, silk, white alloy

女士披巾

乌德穆尔特人 | 19 世纪末 | 布，捻丝，红布，棉织物，金线

Woman scarf

Udmurts | Late 19th century | Cloth, twisted silk, red cloth, cotton, gold thread

女士节日套装
科米—伊泽米亚人
19 世纪末 ~ 20 世纪初
Woman festive attire
Lzhma Komi
Late 19th – early 20th century

衣服

科米人
19 世纪末 ~ 20 世纪初
丝绸，棉织物，棉线

Clothing

Komis
Late 19th – early 20th century
Silk, cotton, cotton thread

衣服

科米人
20 世纪初
丝绸，棉织物，玻璃

Clothing

Komis
Early 20th century
Silk, cotton, glass

女士腰带

科米人，彼尔米亚克
20 世纪中叶
羊毛线

Woman belt

Komis. Permyak
Mid-20th century
Wool thread

围裙

科米人｜ 20 世纪中叶｜棉织物，丝绸，丝带

Apron

Komis｜Mid-20th century｜Cotton, silk, ribbons

女士头饰

科米—济良卡人，东欧，
阿尔汉格尔斯克地区，
梅津斯卡亚地区
20 世纪初
羊毛，锦缎，丝绸，金属
混合纱，玻璃，黄铜（小
亮片），珍珠，陶瓷

Woman headwear

Komi Zyrians.
Eastern Europe,
Arkhangelsk Province,
Mezensky Uyezd
Early 20th century
Wool, brocade, silk, metallic
combination yarn, glass, brass
(sequins), pearls, porcelain

珠串项链

科米—济良卡人，东欧，阿尔汉格尔
斯克地区，梅津斯卡亚地区
19 世纪末
玻璃珠，线（植物纤维）

Beads

Komi Zyrians. Eastern Europe,
Arkhangelsk Province, Mezensky Uyezd
Late 19th century
Glass (beads), thread (plant fiber)

239

披肩

科米—济良卡人，东欧，阿尔汉格尔斯克地区，梅津斯卡亚地区 | 19 世纪末 ~ 20 世纪初 | 丝绸，丝线

Kerchief

Komi Zyrians. Eastern Europe, Arkhangelsk Province, Mezensky Uyezd | Late 19th - early 20th century | Silk, silk thread

女士头饰局部
马里人，东欧，喀山地区，
查雷沃科克沙伊地区
19 世纪末 ~ 20 世纪初
布，棉织物，棉线，丝线，羊毛

Part of a woman headwear
Mari. Eastern Europe,
Kazan Province,
Tsaryovokokshaysky Uyezd
Late 19th – early 20th century
Cloth, cotton, cotton thread,
silk thread, wool

241

女士头饰
马里人，东欧，维亚特卡地区，乌尔汝姆地区 | 19 世纪末 | 布，棉织物，金银饰带，丝线，羊毛线，铜，玻璃珠

Woman headwear
Mari. Eastern Europe, Vyatka Province, Urzhumsky Uyezd | Late 19th century | Cloth, cotton, passementerie, silk thread, wool thread, copper, glass (seed beads)

女士婚礼头巾

马里人，东欧，别尔姆地区，克拉斯诺乌菲姆斯克地区 | 19 世纪末 ~ 20 世纪初 | 布，丝线，羊毛线，黄色合金

Woman wedding head kerchief

Mari. Eastern Europe, Perm Province, Krasnoufimsky Uyezd | Late 19th – early 20th century | Cloth, silk thread, wool thread, yellow alloy

女士头巾

马里人，东欧，乌法省，比尔斯克地区 ｜ 19 世纪末 ~ 20 世纪初 ｜ 布，棉织物，丝线，棉线，白色小亮片

Woman head kerchief

Mari. Eastern Europe, Ufa Province, Birsky Uyezd ｜ Late 19th - early 20th century ｜ Cloth, cotton, silk thread, cotton thread, white sequins

毛巾

马里人，东欧

19 世纪末 ~ 20 世纪初

布，棉织物，羊毛线，丝线，玻璃珠，

白色合金（坠饰）

Towel

Mari. Eastern Europe

Late 19th – early 20th century

Cloth, cotton, wool thread, silk thread,

glass (seed beads), white alloy (pendants)

毛巾

马里人，东欧 | 19 世纪末 ~ 20 世纪初 | 布，棉织物，羊毛线，丝线，玻璃珠，白色合金（坠饰）

Towel

Mari. Eastern Europe | Late 19th – early 20th century | Cloth, cotton, wool thread, silk thread, glass (seed beads), white alloy (pendants)

女士腰带挂饰

马里人，东欧，喀山地区，
科斯莫杰米扬斯克地区
19 世纪末
布，丝带，玻璃珠

Woman belt pendant

Mari. Eastern Europe, Kazan Province,
Kozmodemyansky Uyezd
Late 19th century
Cloth, silk (ribbons), glass (seed beads)

女士腰带坠饰

马里人，东欧，喀山地区，科斯莫杰米扬斯克地区 | 19 世纪末 | 丝带，丝线，亚麻布，丝绳，玻璃珠

Woman belt pendant

Mari. Eastern Europe, Kazan Province, Kozmodemyansky Uyezd | Late 19th century | Ribbons, silk thread, linen, silk cord, glass (seed beads)

7

东欧和
东南欧民族传统
文化中的丝绸

Silk in the
Traditional Culture of
the Peoples of Eastern
and South-Eastern Europe

房屋前的妇女
保加利亚人，东欧，塔夫利亚地区，费奥多西亚地区
1905年
Women in front of the house
Bulgarians. Eastern Europe, Taurida Province,
Feodosiysky Uyezd
1905

年轻的女士们
摩尔多瓦人，东欧，叶卡捷琳诺斯拉夫地区，
巴赫穆特地区
1909年
Group of young women
Moldovans. Eastern Europe, Yekaterinoslav
Governorate, Bakhmut Uyezd
1909

波德涅普罗维耶和波多利亚地区居民传统文化中的丝绸
Silk in the Traditional Culture of Podniprovie and Podolia

19 至 20 世纪初，基辅、波尔塔瓦、波多利亚和切尔尼戈夫等省出现了小型养蚕农场。然而，由于无法大规模种植桑树，这种生产方式并未在该地区普及。

大多数情况下，用于室内装饰和服装的丝织品都是从欧洲或者俄罗斯帝国的其他省份进口，这增加了它们的成本。富裕的人买得起大块的丝绸面料，而不太富裕的阶层则满足于在节日盛装中使用少量的丝绸元素：丝带、织带以及丝线刺绣。女性的头饰用丝绸装饰：缎带装饰少女的花环，装饰已婚妇女的奥奇波克。人们用绸缎缝制女性节日服装——背心、无袖上衣，用丝线刺绣装饰头巾、披肩、衬衫和毛巾。

In the 19th - early 20th centuries, small sericulture farms appeared in the Kiev, Poltava, Podolia, and Chernihiv Governorates. However, this production type did not become widespread in the region, in particular due to the impossibility of large-scale cultivation of mulberries.

In their majority, silk fabrics used for interior decor items and clothing were still imported from Europe or other provinces of the Russian Empire, which increased their cost. Wealthy people could afford larger silk pieces, while less affluent population groups had to settle for small silk elements in festive costumes: ribbons, webbing, and silk thread embroidery. Women's headdresses were decorated with silk: satin ribbons adorned maidens' wreaths, decorated ochipoks of married women. Women's festive clothes were made of satin: vests, sleeveless jackets, while silk thread embroidery decorated headscarves, plakhta fabric, shirts and towels.

斯卢茨克腰带

乌克兰人，东欧
18 世纪 60 ~ 80 年代
丝绸，金线

Kontush (Slutsk) sash

Ukrainians. Eastern Europe
1760–1780s
Silk, gold thread

妇女的毛巾状头饰

乌克兰人，东欧，基辅地区 | 19 世纪末 | 细纱，生丝

Woman towel-like headwear

Ukrainians. Eastern Europe, Kiev Uyezd | Late 19th century | Mousseline, raw silk

无袖外套

乌克兰人，东欧，库尔斯克地区，格拉伊沃隆地区

19 世纪末 ~ 20 世纪初

棉织物，丝绸，厚丝

Sleeveless jacket

Ukrainians. Eastern Europe, Kursk Province,
Grayvoronsky Uyezd

Late 19th – early 20th century

Cotton, silk, heavy silk

250

裹裙

乌克兰人 ｜ 19 世纪 ｜ 乌克兰手工织物，丝线

Warp skirt

Ukrainians ｜ 19th century ｜ Ukrainian handmade fabrics, silk thread

比萨拉比亚地区的蚕丝养殖和丝绸织造
Sericulture and Silk Weaving in Bessarabia

比萨拉比亚养蚕业发展的一个关键因素是 19 世纪初保加利亚人移居此地，他们也形成了自己的养蚕传统。该地区优越的自然环境有利于桑树的种植，这为养蚕业奠定了重要基础。少量本地生丝由农民自己加工，其余的则出售到邻近城镇。在农村地区，半丝绸织物是生产重点：使用生丝线作为纬线，以棉布为基础。传统的女性头巾和家用纺织品就是用这种织物制成的。

A key factor in the development of sericulture in Bessarabia was the resettlement of Bulgarians in the early 19th century, who had developed sericulture traditions themselves. Favorable natural conditions of the region contributed to the cultivation of mulberry trees, which formed an important basis for sericulture. A small amount of raw silk obtained at home was processed by the peasants themselves, the rest went for sale to neighboring cities. In rural areas, semi-silk fabrics were the main focus of the production, raw silk thread was used as weft, and cotton was the basis. Traditional women's headscarves, as well as domestic textiles, were made of such fabric.

织物样品

加告兹人，东欧，比萨拉比亚地区，本德尔地区

19 世纪末

棉织物，生丝

Fabric sample

Gagauz. Eastern Europe, Bessarabia Province,
Bendersky Uyezd

Late 19th century

Cotton, raw silk

织物样品

保加利亚人，东欧，比萨拉比亚地区，伊兹马伊洛夫地区｜20 世纪初｜丝绸

Fabric sample

Bulgarians, Eastern Europe, Bessarabia Province, Izmailsky Uyesd｜Early 20th century｜Silk

枕套

保加利亚人，加告兹人，东欧，
比萨拉比亚地区，本德尔地区
19 世纪末 ~ 20 世纪初
丝绸

Pillow case

Bulgarians, Gagauz. Eastern Europe,
Bessarabia Province, Bendersky Uyezd
Late 19th – early 20th century
Silk

腰带

摩尔多瓦人，东欧，波多利亚省，奥尔戈波利地区 ｜ 1875 ~ 1900 年 ｜ 丝绸

Sash

Moldovans. Eastern Europe, Podolia Province, Olgopol uyezd ｜ 1875-1900 ｜ Silk

254

男式长袍

摩尔多瓦人，东欧，比萨拉比亚地区，基希涅夫地区 | 19 世纪末 ~ 20 世纪初 | 丝绸，棉织物

Male robe

Moldovans. Eastern Europe, Bessarabia Province, Kishinyovsky Uyezd | Late 19th – early 20th century | Silk, cotton

俄罗斯西部各州文化中的丝绸
Silk in the Culture of the Western Provinces of the Russian

　　18 世纪，丝绸服装是波兰最高贵族身份的重要标志。19 世纪上半叶，丝绸进口的低关税和靠近欧洲市场的条件促进了丝绸的传播。20 世纪初，丝绸自产开始活跃。华沙省出现纱线、丝绸漂染和丝带制作等生产业。彼得罗科沃州则生产丝绸和半丝绸织物，包括长毛绒。

　　此外，该地区还形成了几类地方传统服饰的组合，每种组合都以使用丝绸材料为特点。人们用采购的厚天鹅绒、丝绸和绸缎制作无袖外套、衬衫和围裙，这些都是乡村节日套装中的元素。

In the 18th century, silk clothing was an important marker demonstrating the status of the highest Polish aristocracy. Later, in the first half of the 19th century, low duties on imported silk and the proximity of European markets contributed to its propagation. In the early 20th century, the domestic silk production was kick-started. In the Warsaw Governorate, there were yarn production, silk bleaching, dyeing, and ribbon-making productions. The Petrovichi produced silk and semi-silk fabrics, including plush.

In addition, several local traditional costume sets were common across the region, each was characterized by the use of silk materials. Sleeveless jackets, skirts, and aprons were made of purchased fabrics like thick velvet, silk, or satin, which were parts of festive attire in the village.

256

女士套装
波兰人
19 世纪末 ～ 20 世纪初
Woman attire
Poles
Late 19th – early 20th century

披肩

波兰人，东欧，凯莱卡地区，奥尔库什

19 世纪末 ~ 20 世纪初

棉织物

Kerchief

Poles, Eastern Europe,

Kielce Province, Orkush

Late 19th – early 20th century

Cotton

波兰人，东欧，凯莱卡地区，奥尔库什 | 19 世纪末 ~ 20 世纪初 | 羊毛

Eastern Europe, Kielce Province, Orkush | Late 19th – early 20th century | Wool

披肩

Kerchief

无袖上衣

波兰人，东欧，凯莱卡地区，奥尔库什

19 世纪末 ~ 20 世纪初

缎，棉织物，羊毛

Sleeveless jacket

Poles. Eastern Europe, Kielce Province, Orkush

Late 19th – early 20th century

Satin, cotton, wool

围裙

波兰人，东欧，凯莱卡地区，奥尔库什 ｜ 19 世纪末 ~ 20 世纪初 ｜ 羊毛

Apron

Poles. Eastern Europe, Kielce Province, Orkush ｜ Late 19th – early 20th century ｜ Wool

衬衫

波兰人，东欧，凯莱卡地区，
奥尔库什
19 世纪末 ~ 20 世纪初
棉织物，棉线

Shirt

Poles. Eastern Europe, Kielce
Province, Orkush
Late 19th – early 20th century
Cotton, cotton thread

短裙

波兰人，东欧，凯莱卡地区，奥尔库什 | 19 世纪末 ~ 20 世纪初 | 羊毛，棉织物，印花布

Skirt

Poles. Eastern Europe, Kielce Province, Orkush | Late 19th – early 20th century | Wool, cotton, calico

珠串项链

波兰人，东欧，华沙地区
19 世纪 30 ～ 60 年代
玻璃珠

Strings of beads

Poles. Eastern Europe, Warsaw Province
1830-1860s
Glass (beeds)

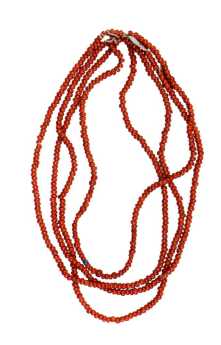

女士头饰

波兰人，东欧，凯莱卡地区，奥尔库什 | 20 世纪初 | 薄纱织物，丝绸，羊毛，金属线，塑料

Woman headwear

Poles. Eastern Europe, Kielce Province, Orkush | Early 20th century | Gauzy fabric, silk, wool, metal thread, plastic

少女头饰

波兰人，东欧，凯尔采地区，米霍夫地区
19 世纪 30 ～ 60 年代
丝绸，金属线

Girl's headwear

Poles. Eastern Europe, Kielce Province,
Miechów Uyezd
1830–1860s
Silk, metal thread

261

儿童帽子

波兰人，东欧，华沙地区，洛维茨地区 | 20 世纪初 | 丝绸，细纱，棉织物

Child cap

Poles. Eastern Europe, Warsaw Province, Łowicz Uyezd | Early 20th century | Silk, mousseline, cotton

结语
Conclusion

　　当历史的长河回溯至19至20世纪，通过丝绸之路传播的欧亚大陆各民族服饰文化无疑是璀璨夺目的。丝绸之路东起中国，西达欧洲，连接着欧亚大陆两端。无论是受到中国丝绸风格影响的突厥-蒙古族服饰和佛教丝织用品，还是在与欧洲的贸易往来中逐渐将巴洛克法式玫瑰作为主要服饰纹样的高加索西南部地区，丝绸之路其实并不是一条明确的路，它只是一个通道，或是一个交流带。

　　我们希望通过这个展览，以丝织制品为媒介，呈现不同地区在政治、经济因素的影响下，以民族文化为纽带，通过接触外来商品而激发艺术创造力，最终形成的被视为民族文化象征的艺术瑰宝。它们的存在，已然成为人们当下探索"丝绸之路"的一份重要历史资料，同时也映射出一个独特的时代印记——那时丝绸像黄金一样灿烂。

When the long river of history goes back to the 19th and 20th centuries, the clothing culture of various ethnic groups in Eurasia spread through the Silk Road is undoubtedly dazzling. The Silk Road starts from China in the east and reaches Europe in the west, connecting the two ends of the Eurasian continent. Whether it is the Turkic-Mongolian clothing and Buddhist silk products influenced by the Chinese silk style, or the southwestern Caucasus region that gradually used the Baroque French rose as the main clothing pattern in trade with Europe, the Silk Road is not actually a well-defined road, it is just a channel or a communication belt.

Through this exhibition, we hope to use silk products as a medium to present the formation process of artistic treasures that are regarded as symbols of national culture. They were formed under the influence of political and economic factors in different regions, with national culture as a bond, and through contact with foreign goods, which stimulated artistic creativity. Their existence has become an important historical material for people to explore the Silk Road, and it also reflects a unique mark of the times that silk was as brilliant as gold at that time.

БОЖЕ ЦАРЯ ХРАНИ